Electric Experiments
for Technology
Second Edition

by Sid Antoch

Electric Experiments for Technology

ISBN-13: 978-1-935422-12-9
ISBN-10: 1-935422-12-X

Cover Pictures

1.   Earth, western hemisphere: NASA
2.   Solar panels: Nellis Air Force Base
3.   http://public domain pictures.net
4.   Portland Community College

Trademark Information

TEKTRONIX,TEK, and OpenChoice™ Desktop
are registered trademarks of Tektronix, Inc.

LTspice®, Linear Technology®
are registered trademarks of Linear Technology.

Microsoft®, Windows®, Excel®, and Word®
are registered trademarks of Microsoft Corporation.

Texas Instruments® and TI®
are registered trademarks of Texas Instruments Corporation.

ZAP Studio LLC
PO Box 1150
Philomath, OR 97370
www.zapstudio.com

Acknowledgement

Dan Kruger, electrical engineering and electrical engineering technology instructor at Portland Community College, made valuable contributions to this second edition. He applied instructor and student feedback from the first edition experiments to correct and clarify the experiments in the second edition.

The Portland Community College Electronics Engineering Technology Program, in Portland, Oregon, and the following instructors, also influenced the creation of the experiments in this lab book: Doug Draper, Mike Farrell, Gary Hecht, Trung Le, and Sanda Williams.

Introduction

Sid Antoch has taught electrical engineering circuits courses at Portland State University and Portland Community College and electrical engineering technology courses at Portland Community College and Tektronix. This lab book is in part the result of this experience.

Experiments in this manual are intended for the laboratory component of an electrical engineering technology electric circuits course. The experiments emphasize the use of spreadsheets and simulation software. Several of the experiments refer to the specific application of a Tektronix TDS type oscilloscope to acquire experimental data. These may be omitted or modified if the lab does not have TDS type oscilloscopes.

Analysis suggestions are provided for each experiment. These may be used as is or may be modified by the instructor according to the lab curriculum requirements. Experiments may be selected according to the accompanying textbook and course emphasis.

LTspice is used exclusively in the second edition because it was found to be easier for students to download and use. More emphasis is placed on single phase and 3-phase power distribution, including distribution losses and power factor compensation.

Most of the labs may be performed using a solder-less breadboard and standard lab equipment. The 3-phase experiments require a low voltage source of 3-phase power. Several options are available. A "Phase Tripler" circuit is presented which students may build on a breadboard. Also a bench top 3-phase voltage source is available from ZAP Studio Electronics as well as an inexpensive circuit board which may be plugged directly into a breadboard. These are described in the appendix.

Equipment List

Digital Oscilloscope with 10X probes, Function generator, Digital Multi-meter.

Solder-less Breadboard. Power Supply: 0-volts to 6-volts and 0-volts to ±15-volts.

Phase Tripler (refer to appendix 2). Motor-Generator Set (refer to appendix 3).

Computer with word processor, spreadsheet, simulation, and data acquisition software.

Parts List

Resistors: 8.2, 10, 47, four 100, 150, 390, 470, three 220, 330, Three 680, 750, two 1K, 1.2K.

Resistors: two 1.5K, 1.8K, 2.2K, 3.3K. 3.9K, 4.7K, 6.8K, 8.2K, five 10K, 15K, 18K, 22K, 33K.

Resistors: 100K, 1Meg, 2.2Meg. All resistors are ¼ watt, 5%.

Power resistor: 10-ohms, 10-watts, 5%.

Trimmer Potentiometers: 100, 5K, 10K, 5%, single turn to fit breadboard.

Capacitors: 1nF, 10nF, 22nF, 47nF, 100nF, 470nF, all 5%, 50-volt minimum.

Capacitors: 1µF, 2.2µF, 4.7µF non-polarized, all 5%, 50-volt minimum.

Inductor: 10mH, 5% (air core). (J.W. Miller 70F102AI).

Inductors: Three 100mH, 5% (iron core, 50mA minimum, 120-ohm resistance maximum).

Light Emitting Diode (LED): Red (other colors may be used).

Transformer: Step-down, 120VAC to 12VAC, about 1-amp, center-tapped.

Audio Output Transformer: 1000-ohms, center tapped, to 8 ohms, 200mW.

Loudspeaker: 8-ohm, 200mW minimum, (2 to 4 inch).

Batteries: Two alkaline batteries (AA 1.5-volt). 6-volt, 4amp-hr, sealed lead-acid battery.

Flashlight bulb and socket: 2.5-volt, 300mA (#14).

On-off switch: (SPST type, toggle or slide).

Incandescent bulb: Type 2182 (14-volt, 80mA).

Analog meter: 1mA < 1000-ohm (or 100µA < 10,000-ohms).

Transistor: 2N3904 or 2N2222

Motor-generator circuit (refer to Appendix 3).

Phase Tripler (refer to Appendix 2).

Contents

	Experiment	Page
1	Electrical Resistance and the Resistor	1
2	Flashlight Circuit Measurement and Simulation	6
3	LTspice Simulation of the Flashlight Circuit	9
4	Power Supplies and Batteries	12
5	Voltage, Current, and Power	16
6	Analog "D'Arsonval" Meter	20
7	DC Measurements and Meter Loading	23
8	LTspice Circuit Simulation	26
9	Kirchhoff's Voltage and Current Laws	28
10	Potentiometer Voltage Dividers	31
11	Thevenin's Theorem and the Bridge Circuit	35
12	Transistor / Dependent Current Source	40
13	Mesh Current Analysis	44
14	Node Voltage Analysis	46
15	Network Theorems	49
16	Superposition and the Voltage Translator	52
17	Function Generator and Oscilloscope	54
18	RC Transient Response	61
19	RL Transient Response	65
20	Capacitor Network Transient Response	69
21	Superposition of AC and DC Voltages	71
22	Working with Phasors	74
23	AC Measurements / Series RC Circuit	76
24	AC Measurements / Series RL Circuit	80
25	Series-Parallel AC Circuit Measurements	82
26:	Capacitive Voltage Dividers	85
27:	Two-Source AC Circuit	88

28:	Thevenin's and Norton's Theorems	90
29:	DC Motor and Generator Basics	92
30:	AC Power Basics	96
31:	2-Phase Power Distribution	99
32:	Power Factor Compensation / Parallel Circuit	103
33:	Series Compensation and Power Transfer	106
34:	Low-Pass Filters	109
35:	High-Pass Filters	112
36:	Audio Crossover Network	115
37:	Two-Pole Low-Pass and Band-Pass Filters	118
38:	Series Resonant Passive Band-Pass Filter	122
39:	Parallel Resonant Band-Pass Filter	126
40:	Audio Output Transformer	129
41:	Three Phase Power / Wye Connection	133
42:	Three Phase Power / Delta Connection	137
43:	Fourier Series and Circuit Analysis	140
44:	Band-Pass Filter / FFT / Square Wave	144
45:	Band-Pass Filter / FFT / Triangle Wave	148
Appendix 1: Electric Circuits Lab Report Information		150
Appendix 2: Phase Tripler Circuit Information		153
Appendix 3: DC Motor-Generator Information		157
Appendix 4: Flashlight Circuit Information		158
Appendix 5: LTspice Information		158

History Notes

Georg Simon Ohm	5	Edward Lawry Norton	48
Count Volta	11	Michael Faraday	53
André-Marie Ampère	15	Joseph Henry	68
Gustav Robert Kirchhoff	30	James Clerk Maxwell	79
Léon Charles Thévenin	39	Sir Charles Wheatstone	102
Thomas Alva Edison	43	Jean Baptiste Joseph Fourier	143

Experiment 1: Electrical Resistance and the Resistor

Introduction

Ohm's law is the most fundamental equation in electric circuit analysis. It states that the amount of electric current flowing in a circuit is directly proportional to the voltage applied to the circuit, and inversely proportional to its resistance.

$$I = \frac{V}{R}$$ I is current in amperes, V is potential difference in volts, and R is resistance in ohms.

Resistors are used in electric circuits to control the flow of current. Resistors are commercially available which have a specific amount of resistance and power dissipation ability. The amount of resistance is usually marked on the resistor using a color code. The power dissipation is determined by the physical size of the resistor.

An "ohmmeter" is used to measure resistance. Most ohmmeters are part of an instrument that is also capable of measuring other electrical quantities, such as voltage and current. These are typically called "multi-meters", and since they usually have a digital display, they are called "digital multi-meters" or "DMMs" for short. To use the DMM you need to know how to set it to make the desired measurement (function), and how to set it for best accuracy (range).

Objectives

The purpose of this lab exercise is to learn how to measure resistance with the DMM. An error analysis will compare the measured resistor values to the labeled resistor values using a spreadsheet. In addition, you will measure the resistance of series and parallel combinations of resistors, and compare the results to theoretical calculations based on equations provided.

Series and parallel connections will be made using a solder-less breadboard. The object of this part of the exercise is to learn to use the breadboard. *Theoretical knowledge of series and parallel resistor connections is not expected.*

Resistance values are read using the color code given below.

Standard Resistor Color Code

Color	Value	Color	Value
Black	0	Blue	6
Brown	1	Violet	7
Red	2	Gray	8
Orange	3	White	9
Yellow	4	Gold	0.1 / 5%
Green	5	Silver	0.01 / 10%

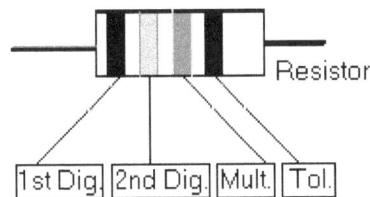

1st Dig. | 2nd Dig. | Mult. | Tol.

Resistor

This color code is for "standard" resistors with an accuracy rating, or "tolerance", of ±5% or ±10%. That is, their value is guaranteed to be within ±5% or ±10% of their labeled value.

Their colors are read from left to right. The first two color bands represent the first two significant digits of the resistor value. The color of the third band represents a multiplier of 10^N, where N is the value represented by the color.

The fourth band is always gold or silver, which indicates a tolerance of $\pm 5\%$ or $\pm 10\%$. The first band is never gold or silver. *So to read a resistor's value correctly, the gold or silver band must be on the right.*

For example, a resistor whose first band is red, second band is yellow, third band is orange, and fourth band is gold, has a value of 24,000 ohms (24×10^3), and a tolerance of 5%.

Resistance Measurement

A digital multi-meter (DMM) will be used for measuring the resistance values. The instructor may explain the operation of the instrument before you use it for the first time. You may also check to see if an instruction manual is available for the instrument. The DMM may have buttons and/or switches to its function and range.

Set the function to "OHMS". Some meters are capable of automatically setting the range to get the most accurate reading, which is related to the number of significant digits displayed. You should be able to get at least three significant digits of accuracy. Experiment with the range settings when making the measurements specified in the procedure below.

The power rating of each resistor is determined by its physical size. Smaller dimensions represent a smaller power handling capability. A sample of several different size resistors should be available in the lab. A very common power rating is ¼ watt. If a ¼ watt resistor dissipates more than ¼ watt it will get excessively hot and may burn out.

Procedure

Equipment and Parts
DMM and Breadboard. Resistors: 1K, 4.7K, 10K, ¼ watt, 5% or 10% tolerance.

Part A: Measurements and the Spreadsheet

Do not touch the metal tips of the DMM probes when making measurements.

1. Use the resistor color code to select the 1K, 4.7K, and 10K resistors. Determine their tolerance. Measure the values of the resistors with the DMM to at least three significant digits.

2. Enter the results into a spreadsheet. Calculate the deviation of each resistor's measured value compared to its labeled value. Calculate the percent deviation of each resistor's measured value from its labeled value. Refer to the example on the next page. Note that starting a spreadsheet's cell with an equal sign indicates the cell contains a formula.

Use the spreadsheet layout shown below to do the calculations.

	A	B	C	D	E	F
1	Resistor	Labeled Val.	Measured Val.	Deviation	Labeled %	Measured %
2	1	1000				
3	2	4700				
4	3	10000				
5						
6						

Deviation: =C2-B2   Percent deviation: =(D2/B2)*100
Enter the expression for deviation into cell D2 and percent deviation into cell E2. Use the "fill down" feature of the spreadsheet to calculate rows 3 and 4.

Part B: Series and Parallel Connections

Before starting this exercise (and the other exercises in this manual) you need to have a way of connecting electronic parts together into a circuit. An easy and very common method to quickly connect parts together is to use the "solder-less breadboard", also called a "prototyping board" or "protoboard". The board has holes 0.1 inches apart into which component leads can be inserted.

Solder-less breadboards are available from a variety of manufacturers and sources, in a variety of sizes, but they all have the same arrangement of the holes and connections

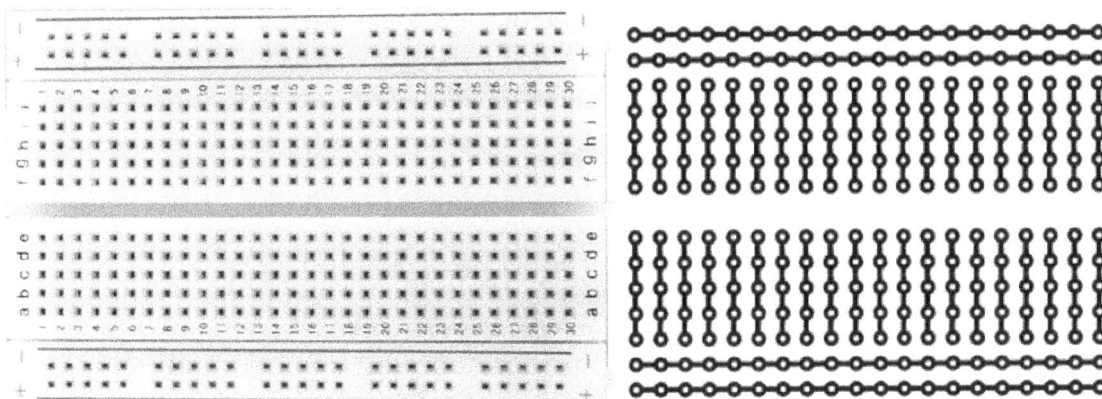

The picture above on the left shows a typical breadboard. Components such as resistors, capacitors, transistors, integrated circuits, and wires can be plugged into it. The picture above on the right shows how the holes are connected. You should memorize these connections.

Components such as resistors can be connected in series, parallel, and in a combination of series and parallel. The following exercises show how to connect resistors in series and parallel, and how to measure the resistance of the series and parallel combinations.

The measurements will be compared to the theoretically expected values using the equations provided. If a measurement does not agree with the calculation, check the breadboard connections and the labeled values of the resistors.

1. Connect your 1K, 4.7K and 10K resistors in series. Measure the resistance, R_{TS}, of the combination as shown in the circuit's schematic diagram below. A connection example is shown on the right.

Ohmmeter leads connect between points a and b.

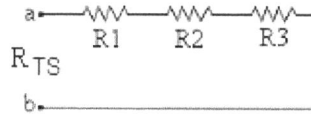

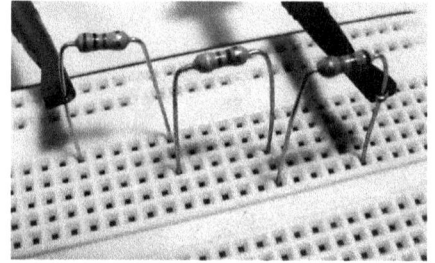

R_{TS} = _____ (measured total series resistance)

2. Calculate the theoretical resistance of this series combination as given the equation:

$$R_{TS} = R1 + R2 + R3.$$

Use the measured values of the resistors from part A. Enter the equation into your spreadsheet and have the spreadsheet do the calculation. Also enter the measured value into the spreadsheet as shown in the example spreadsheet on the next page.

3. Connect the 1K, 4.7K and 10K resistors in parallel as shown in the diagram below and picture on the right. Measure the resistance, R_{TP}, of the parallel combination.

Ohmmeter leads are connected between points a and b.

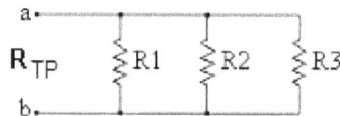

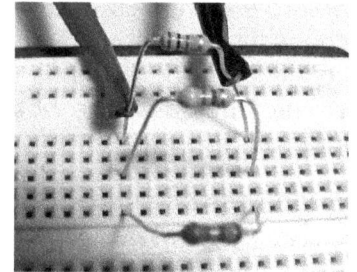

R_{TP} = _____ (measured total parallel resistance)

4. Calculate the resistance of this parallel combination using the equation below.

$$\frac{1}{R_{TP}} = \frac{1}{R1} + \frac{1}{R2} + \frac{1}{R3} \quad \text{so that} \quad R_{TP} = \frac{1}{\left(\dfrac{1}{R1} + \dfrac{1}{R2} + \dfrac{1}{R3}\right)}$$

Use the measured values of the resistors from part A. Enter the equation into your spreadsheet and have the spreadsheet do the calculation. Also enter the measured value of the series resistance and the measured value of parallel resistance into the spreadsheet as shown in the example spreadsheet below.

	A	B	C	D	E	F
1	Resistor	Labeled Val.	Measured Val.	Deviation	Labeled %	Measured %
2	1	1000	996	-4	0.05	-0.4
3	2	4700	4760	60	0.05	1.276595745
4	3	10000	9720	-280	0.05	-2.8
5						
6						
7		Calculated Val.	Measured Val.	Deviation	% Deviation	
8	Series	15476	15390	-86	-0.555699147	
9	Parallel	759.3125376	762.2	2.8874624	0.380273246	

Equation in B8: =C2+C3+C4   Equation in B9: =1/(1/C2+1/C3+1/C4)

4

LAB REPORT

1.   Open a word processor document and save it as: "Experiment 1 Report". Use the following format:

Student name and lab partner name (if applicable).
Course number: Lab experiment number and name.

Example:

George Jones and Sally Smith

EET111: EXPERIMENT 1: Electrical Resistance and the Resistor

2.   Copy your spreadsheet results and paste them into the document.

The instructor will specify how to turn in the report. You may just need to show the resulting document on the lab computer, or the instructor may also ask for additional analysis, including a more comprehensive lab report.

History Note:

Georg Simon Ohm (17 March 1789 – 6 July 1854) was a German physicist. As a high school teacher, Ohm began his research with the recently invented electrochemical cell, invented by Italian Count Alessandro Volta. Using equipment of his own creation, Ohm determined that there is a direct proportionality between the potential difference (voltage) applied across a conductor and the resultant electric current. This relationship is now known as Ohm's law.

From Wikipedia, the free encyclopedia

Experiment 2: Flashlight Circuit Measurement and Simulation

Introduction

Batteries are a common source of stored electrical energy. Batteries are available in a wide variety of physical sizes, voltage ratings, and capacity ratings.

The voltage rating relates the battery's ability to cause electric current to flow in an electric circuit. Ohm's law states that electric current is directly proportional to voltage.

Capacity rating is usually expressed in units of "amp-hours". It is the product of the battery current in amperes and the length of time the current is delivered in hours. The capacity rating of a battery depends on the battery type and on its physical size. It also depends on the discharge rate (how fast or slow the battery is discharged).

Another battery property is "internal resistance". A battery's internal resistance reduces the amount of energy that can be extracted from a battery. Some of the battery's voltage is dropped across the battery's internal resistance resulting in lower battery output voltage.

Energy is dissipated by the battery's internal resistance as heat. The energy dissipated is proportional to the square of the battery's output current, so higher currents will decrease the energy (amp-hours) obtainable from the battery. A simple model of a battery is shown below.

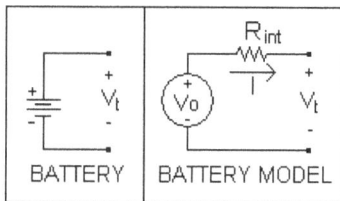

Vo in the battery model represents an "ideal" voltage source. The battery terminal voltage, V_t, is equal to Vo when no current is drawn from the battery (V_t = Vo).

This is referred to as the "open circuit" voltage. V_t is given by the equation: $V_t = Vo - I R_{int}$, where I is the current in amperes.

A very common non-rechargeable battery is the alkaline type. The alkaline battery is typically used in flashlights, portable radios, and other portable electronics. Two alkaline batteries will be used in this lab exercise.

Objectives

The objectives of this lab include an introduction to the battery as a voltage source, an introduction to voltage measurement, and an introduction to a simple flashlight circuit. Circuit simulation software will be used to simulate this lab exercise in experiment 3.

Procedure

Equipment and Parts

Digital Multi-meter, two alkaline batteries (AA 1.5V).
Flashlight bulb and socket, 2.5V, 300mA (#14).
On-off switch (SPST type, toggle or slide).
Computer with circuit simulation software
(Appendix 4 describes this flashlight circuit built on a pine board)

Observe the following precaution before proceeding with this exercise:
Do not short circuit a battery. Connecting a low resistance across the battery terminals can result in excessive current, which will not only discharge the battery quickly, but it may also cause the battery and the wires to dangerously overheat. Batteries are relatively safe when treated properly.

Flashlight Circuit

1. Your circuit should be connected and labeled as shown on the right. Make sure that the switch is in the off position.

    There are 4 nodes identified for voltage measurements: A, B, C, and Ground.

    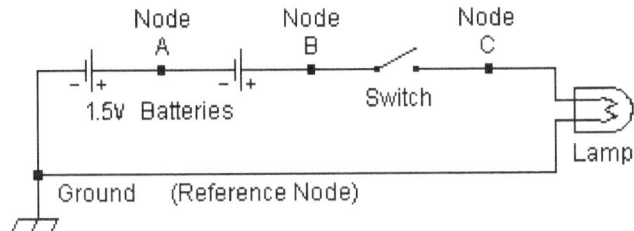

2. Set the digital multi-meter (DMM) to measure DC voltage. Set the "range" for the most accurate measurement (most significant digits). Connect the black lead (negative reference) of the DMM to the circuit's ground node.

    Before proceeding to the next step, be absolutely sure that the DMM is set to measure voltage. The batteries are capable of supplying enough current to damage the DMM if the DMM is set to measure current when it is connected directly to a battery. Also, make sure the voltage across each battery's terminals is approximately 1.5 volts. If the voltage is significantly lower, you may need to replace the battery before proceeding.

3. Make sure that the circuit works by turning the switch on and off (the lamp should go on and off). With the switch off, use the red (positive reference) meter lead to measure and record the voltages: V_A at Node A, V_B at Node B, and V_C at Node C.

    Switch Off: $V_A =$ _____ $V_B =$ _____ $V_C =$ _____

4. Turn the switch on (lamp on) and immediately measure and record the voltages: V'_A at Node A, V'_B at Node B, and V'_C at Node C. The voltages will decrease over time as the batteries are drained. Turn the switch off after making these measurements. The apostrophe simply indicates that, for example, X' is not the same variable as X. X' would be pronounced, "x prime".

    Switch On: $V'_A =$ _____ $V'_B =$ _____ $V'_C =$ _____

5. The battery current can be measured by setting the DMM to measure current. Set the DMM to the 2 ampere current range. Ask the lab instructor for help if you are unsure of how to do this. To measure current, you must break, or open, a connection and reroute the current through the motor.

6. With the flashlight circuit switch off, connect the black lead (negative) of the meter to node C and red lead to node B. The lamp should light and the meter will read the flashlight circuit current, I. Record the current below:

    I =_____ A.

7. You should observe the following:

    a. V_A and V_B decreased between steps 3 and 4.

    b. V_C is 0 when lamp is off, and equal to V_B when lamp is on.

    Consider the battery model on the right. Vo is the unloaded battery voltage, V_B, from step 3. V_t is the loaded battery voltage, V'_B, from step 4. The battery's terminal voltage is given by $V_t = Vo - I\ R_{int}$.

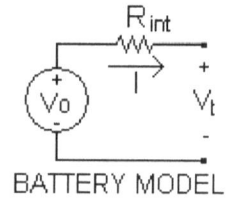

    The battery's internal resistance for the battery model can be calculated from the equation below on the left and for the flashlight circuit using the equation on the right.

Battery Model: $R_{int} = \dfrac{Vo - V_t}{I}$	Flashlight Circuit: $R_{int} = \dfrac{V_B - V'_B}{I}$

Analysis

1. Calculate the internal resistance of your batteries, R_{int}. This is the total resistance of the two batteries in series. using the equation on the right above.

    $R_{int} =$_____ Ω.

2. Calculate the lamp's effective resistance using Ohm's law: $R_L = V'_C / I$.

    $R_L =$_____ Ω.

3. Use the open circuit terminal voltage and the internal resistance of your batteries to calculate the instantaneous current that would flow through the battery if the battery were shorted (0Ω connected across it). Consider the internal battery resistance and Ohm's law.

    $I_{Short} =$_____ A.

4. Calculate the battery life of your flashlight circuit if the batteries are rated at 2 amp-hours for typical flashlight applications.

    Life =_____ Hrs.

8

Experiment 3: LTspice Simulation of the Flashlight Circuit

Introduction

Circuit simulation software is commonly used to test and analyze the operation of circuit designs on a computer. There are two popular simulation programs that may be downloaded for free: OrCAD PSpice evaluation version, and LTspice. All the examples in this lab book use LTspice.

LTspice Circuit Simulation

LTspice IV may be downloaded for free from Linear Technology's website:
http://www.linear.com/designtools/software/

You can also download a manual and a "getting started guide". There are no limits on the schematic size or number of components used. It is easy to use, easy to save your work, and it is used at a number schools and universities.

Drawing the Circuit

1. Start the *LTspice* program. In the main menu bar, click on File and then click on *New Schematic*.

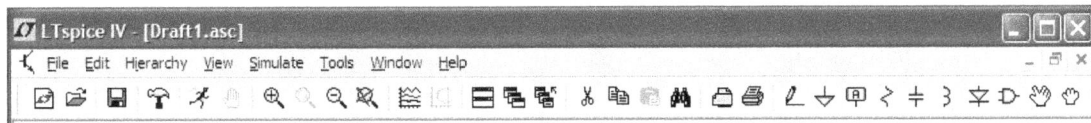

2. Create the new schematic. Left click on the resistor symbol and drag and place the resistor, R1. Repeat to get and place resistor R2.

   Selected components may be rotated by using "CTRL" and "R" keys on the keyboard. Components may be deleted using the "scissors" in the main menu.

   Right click on each "*R*" of each resistor. Change the values to the measured values of the resistances from your experiment.

   R1 is the battery's internal resistance and R2 is the resistance of the light bulb.

3. Left click on the gate symbol between the diode and the hand in the main menu to get the dialog box shown on the right.

   Select the part *voltage* and place it in the schematic. Set the value of V1 to the measured open circuit voltage of your batteries.

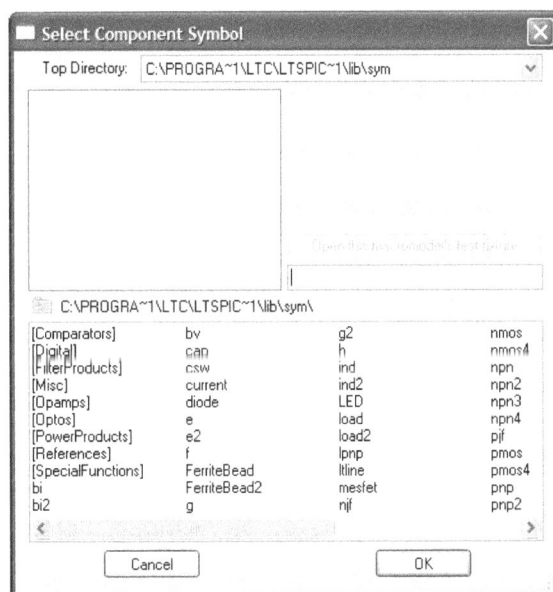

Left click on the ground symbol and place it under V1 as shown on the right.

4. Left click on the "pencil" in the main menu to connect the components.

5. Left click on the "A" between the ground symbol and the resistor symbol in the menu bar to label the nodes *N1, N2,* and *N3* as shown on the schematic. Save your project before going further.

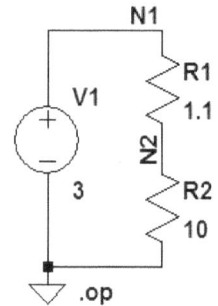

Experiment with some of the other menu items, such as the scissors to delete components and wires, the "hands" on the right side of the menu bar to move and drag components, the "undo" and "rotate" icons to the right of the hands.

6. Click on Simulate in the main menu and select Edit Simulation Cmd. Select DC op pnt. Click on ok. A ".op" command will appear which can be placed anywhere on the schematic.

7. Click on Simulate and select *Run.* If there are no errors, you will see an *"Operating Point"* file as shown below.

The results below show that the lamp current is 0.27A and the lamp voltage is 2.7V. The power dissipated by the lamp is: P = I V = .27(2.7) = 0.73W.

```
* C:\Program Files (x86)\LTC\LTspiceIV\Draft3.asc        [x]

      --- Operating Point ---

V(n1):          3              voltage
V(n2):          2.7027         voltage
I(R2):          0.27027        device_current
I(R1):          0.27027        device_current
I(V1):          -0.27027       device_current
```

Analysis

1. Compare the results of your simulation to your measured values by calculating the percent difference between the measured and simulated values of circuit current and lamp voltage.

2. Use the values of V1, R1, and R2 in your simulation circuit model to calculate the lamp current. Compare the calculated current to the simulated current and the measured current. Express the percent difference between the calculated current and the simulated current. Express the percent difference between the calculated current and the measured current.

3. Lab report option: Open a word processor document. Title it as follows:
   Your Name   Course Number   Date   Experiment Name and Number.

4. Use "copy and paste" to copy the schematic diagram of your circuit simulation into the document. Label the diagram. Use "copy and paste" to copy the output file of your simulation and paste only the analysis results into the document.

5. Label your schematic and simulation results. Express the results of your measurements, calculations, and simulation in a short paragraph.

6. The lab instructor may ask for a printed copy of your document, an emailed copy, or may just ask to see it displayed on the lab computer.

History Note:

Count Alessandro Giuseppe Antonio Anastasio Volta (18 February 1745 – 5 March 1827) was an Italian physicist known especially for the invention of the battery in 1800.

From Wikipedia, the free encyclopedia

Experiment 4: Power Supplies and Batteries

Introduction

Workbench type DC power supplies are used to power experimental circuits in laboratories, as substitute power supplies for circuits on a test bench, as well as for many other applications. The typical bench power supply for general electronics applications will have several variable output voltages.

The typical power supply may have a variable 0 to 20V source with a positive output with respect to ground and a variable 0 to 20V source with a negative output with respect to ground. The maximum current for these may be about 1A. In addition, it may have other sources that may be utilized.

Rechargeable lead-acid batteries are a very common voltage source. This battery is available in a variety of forms: wet cell, sealed lead-acid, and gel cell. The wet cell is typically what you'll find in an automobile. This battery contains sulfuric acid and must be handled carefully.

The sealed lead acid battery is designed to be "maintenance free". It has the advantage of being "non-spill-able", but one must be extra careful when charging it. The gel cell battery has a "gelled" lead-acid electrolyte. When charging any lead-acid battery, one must be careful not to overcharge the battery. Overcharging can result in battery damage and may be dangerous.

Lead-acid batteries are used in cars, boats, and aircraft. They can supply large currents for short periods to start engines. They can be recharged hundreds of times. Lead-Acid batteries are also used to store power from wind generators and solar panels.

Other common rechargeable batteries include the Nickel-Cadmium, Nickel-Metal Hydride, and Lithium-Ion. These batteries are typically used in cell phones and notebook computers.

The table below compares typical characteristics of a variety of battery types. Actual characteristics of specific batteries depend on the manufacturer and technology.

Battery Type	Chargeable	Battery Volts / Size for Specs	Amp-Hrs	Cycles	Comments
Sealed Lead-Acid	Yes	6.6V / 70 mm by 47 mm by 102 mm	4.2	800	No Memory Effect
Nickel-Cadmium	Yes	1.2V / D 32 mm dia. by 62 mm long	2	500	Memory Effect
Mag-Nickel Hydride	Yes	1.2V / D 32 mm dia. by 62 mm long	4	500	Some Memory Effect
Lithium Ion	Yes	3.7V / 18.3 mm dia. by 50 mm long	1.2	500	No Memory Effect
Lithium	No	3.0V / 17 mm dia.by 45 mm long	2.5	1	non-rechargeable
Alkaline	No	1.5V / D 32 mm dia. by 62 mm long	10	1	non-rechargeable

Objectives

This lab is an introduction to DC power supplies and lead-acid rechargeable batteries. The charging and discharging characteristics of a small 6V lead-acid battery will be investigated.

Procedure

Equipment and Parts

DMM, DC Power Supply, and Breadboard.
Resistor: 10Ω, 10W, 5%.
6V, 4A-Hr, sealed lead-acid battery.

Part A: Lead-acid Battery Discharging

Please observe the following precaution before proceeding with this exercise:

Do not short circuit a battery. Connecting a low resistance across the battery terminals can result in excessive current, which will not only discharge the battery quickly, but it may also cause the battery and the wires to dangerously overheat. Batteries are relatively safe when treated properly. As a an extra precaution, a 1A fuse is connected in series with the battery.

The battery is modeled as an ideal source (Vo) in series with its internal resistance (R_{int}) .

1.   Connect the circuit on the right with the jumper, JMPR, open (battery is not connected to the 10Ω load resistor).

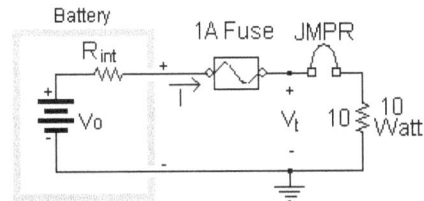

Vo is the open circuit battery voltage. V_t is the battery terminal voltage. R_{int} is the battery's internal resistance.

2.   Measure and record Vo by measuring V_t when the 10Ω load resistor is not connected. V_t = Vo when load is not connected.   Use a battery whose voltage is between 6 and 6.6V.

Vo = _____

When the jumper is connected, the battery will discharge through the load resistor and V_t < Vo. A discharge current of about 0.6A will flow and create a voltage drop   across the battery's internal, R_{int}.

3.   Connect the jumper to the battery at t = 0. Measure and record $V_{t(0)}$. Leave the jumper connected. Measure and record $V_{t(20)}$ at t = 20 minutes. If a stopwatch is not available, the website http://www.online-stopwatch.com/ can be used.

Time	t = 0 minutes, $V_{t(0)}$	t = 20 minutes, $V_{t(20)}$
Voltage V_t		

Disconnect the jumper when done.

Part B: Lead-acid Battery Charging

Please observe the following precaution before proceeding with this exercise:

Be extremely careful to not overcharge a rechargeable battery. It is dangerous to overcharge batteries because the chemical reactions can emit dangerous gasses, chemicals and heat, which may cause the battery to burst or explode.

A 6V battery whose terminal voltage exceeds 6.6V is in danger of being overcharged.

1. Connect the circuit on the right, but do not connect the 10Ω resistor yet. Set Vs to 12.0V using the DMM.

    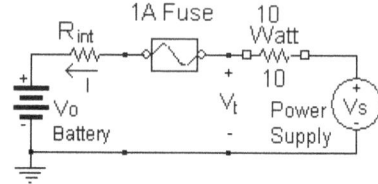

    Vo is the open circuit battery voltage. V_t is the battery terminal voltage. R_{int} is the battery's internal resistance.

2. Measure and record V_t. V_t = Vo in this case because the battery is disconnected.

    Vo = _____

3. Connect the 10W, 10Ω resistor at t = 0. Make sure that Vs is still 12.0V.   Measure and record $V_{t(0)}$ at t = 0 and $V_{t(20)}$ at t = 20 minutes.

    You may consider doing something else while the battery is charging, such as the analysis of part A.   But do not forget to disconnect the battery from the power supply in exactly 20 minutes.

Time	t = 0 minutes, $V_{t(0)}$	T = 20 minutes, $V_{t(20)}$
Voltage V_t		

Disconnect the circuit when done.

Analysis, Part A

1. Calculate R_{int} using the values of Vo and V_t at t = 0.

2. Calculate the average value of V_t:

$$V_{t(ave)} = \frac{V_{t(0)} + V_{t(20)}}{2}$$

3. Calculate the average discharging current, I_{ave}, using the measured value of your 10Ω resistor.

$$I_{ave} = \frac{V_{t(ave)}}{10}$$

4. Use the average values calculated above to calculate the amount of energy delivered to the load in 20 minutes in joules and in watt-hours.

14

5. Calculate the amount of energy dissipated by the battery's internal resistance in joules and in watt-hours during the 20 minutes of discharge.

Analysis, Part B

1. Calculate R_{int} using the values of Vo and V_t at t = 0.

2. Calculate the average value of V_t:

$$V_{t(ave)} = \frac{V_{t(0)} + V_{t(20)}}{2}$$

3. Calculate the average charging current, I_{ave}, using the measured value of your 10 Ω resistor.

$$I_{ave} = \frac{V_{t(ave)}}{10}$$

4. Use the average values calculated above to calculate the amount of energy delivered to the battery in 20 minutes in joules and in watt-hours.

History Note:

André-Marie Ampère (20 January 1775 – 10 June 1836) was a French physicist and mathematician who is generally regarded as one of the main discoverers of electromagnetism. The SI unit of measurement of electric current, the ampere, is named after him.

From Wikipedia, the free encyclopedia

Experiment 5: Voltage, Current, and Power

Introduction

Electrical devices which do work consume energy. Power is the rate that energy is consumed. Equations describing the relationship between power, energy, voltage, and current are given below. P is power in watts, W is work (energy) in joules, I is in amperes, and V is in volts:

$$P = \frac{dW}{dt} \qquad P = IV \qquad V = IR$$

The first equation is a calculus expression stating that power is simply the rate at which energy, or "W", is changing with respect to time, or "t".

All electric power sources are basically sources of potential difference or voltage. Examples include batteries, generators, solar cells, and power supplies.

A typical lab power supply has variable (adjustable) voltages from 0 to about 30V at a maximum current of about 1A. Maximum ratings are often printed on the front panel. Note what maximum voltages and currents are available from your power supply. You may refer to the instrument's manual, if available, for more details.

Voltage Sources

Voltage sources used in circuit analysis are usually "ideal" sources. By ideal we mean that the value of the voltage of the source does not change regardless of how much current it supplies. Practical sources are limited in the amount of current they can supply.

A voltage-regulated power source will supply a constant output voltage, V_S, when the output current, I, is less than its maximum limit, I_{lim}. Some power supplies have an adjustable current limit, I_{lim}, and they will supply a constant output voltage for load currents less than I_{lim}. Refer to the graph below.

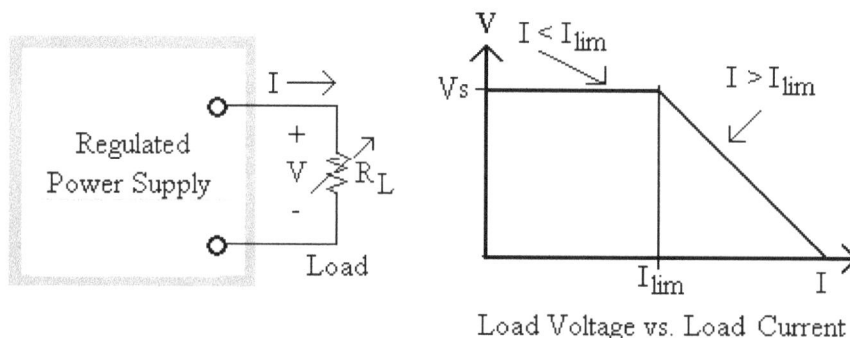

Load Voltage vs. Load Current

A power supply will be used to supply power to a light emitting diode (LED) in part A and to a light bulb in part B. The voltage versus current characteristics and power dissipation of both will be measured and compared.

Voltage Measurement

The DMM will be used to measure voltage. *To measure voltage the meter leads are connected across (in parallel with) the device.* Voltage is measured between two nodes. Voltage is a potential difference. You are measuring the potential difference between two nodes.

Current Measurement

To measure current the meter must be connected in series with the circuit so that the current flowing through the ammeter is the same as the current flowing through the circuit. The circuit must be temporarily disconnected to insert the ammeter in series with it.

Current can also be determined by measuring the voltage drop across a known resistance and calculating the current flowing through it using Ohm's law. Current through an LED and an incandescent lamp will be determined from the voltage measured across a series connected resistor.

Objectives

The voltage versus current graph of a resistor is a straight line. The resistance of a resistor is constant and does not change with the current through it. This experiment will show that this is not the case for LEDs and light bulbs. Objectives of this exercise include gaining experience with voltage measurement, spreadsheet calculations, and spreadsheet graphing.

Procedure

Equipment and Parts

Power supply, DMM, and Breadboard
Resistors: 10Ω and 1000Ω (1/4 watt, 5%).
Light bulb: Type 2182 (14V, 80mA). Light Emitting Diode (LED).

Part A: I versus V for an LED

1. Connect the circuit below. The positive lead of the LED is the anode. It is the longer lead and it should be connected to point A on the diagram below. Turn on the power supply and set the voltage to zero. Set the DMM function to volts and set the range to 20V.

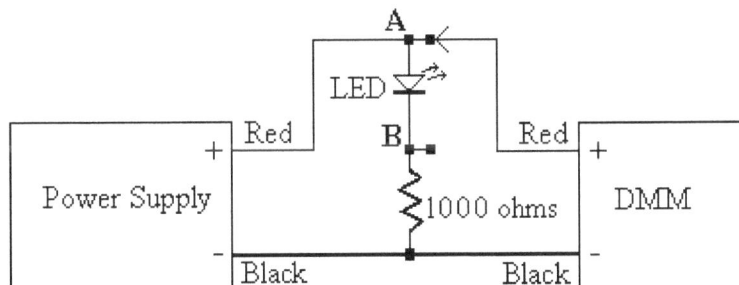

2. The red DMM lead will be moved between points A and B follows:

(a) Set the power supply voltage to exactly 0.0V (as read on the DMM). The DMM's red lead should be connected to point A.

(b) Move the DMM's red lead to point B. Measure and record the voltage at B.

(c) Repeat steps (a) and (b) for voltages at point A) of:
1.0, 2.0, 3.0, 4.0, 5.0, 6.0, 7.0, 8.0, 9.0, 10.0, 11.0, and 12.0V.

Create a spreadsheet table as shown below. You will have 13 rows of data.

	A	B	C	D	E
1	Volts at A	Volts at B	Volts LED	A LED	Watts LED
3	0	0	0	0	0
4	1				
5	2				

C4 has: =A4–B4   D4 has: =B4/1000   E4 has: =C4*D4

You can automatically fill a column of data. For example, select cell C4, then click the mouse on the bottom right corner of the cell, and drag the mouse down to fill all of the rows of that column. This can also be done with copy and paste using ctrl+c and ctrl+v.

Part B: I versus V for a Lamp

1. Connect the circuit below. The lamp is not "polarized". Its leads can be connected in either direction. Make sure that the power supply output voltage is set to zero. Set the DMM function to volts and set the range to 20V.

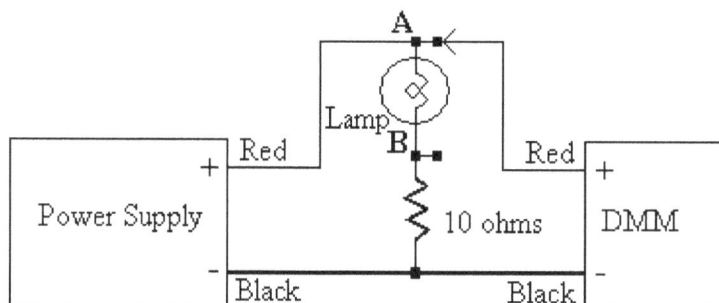

2. The Red DMM lead will be moved between points A and B as follows:

(a) Set the power supply voltage to exactly 0.0V (as read on the DMM). The DMM's red lead should be connected to point A.

(b) Move the DMM's red lead to point B. Measure and record the voltage at B.

(c) Repeat steps (a) and (b) for voltages (as read by the DMM at point A) of:
1.0, 2.0, 3.0, 4.0, 5.0, 6.0, 7.0, 8.0, 9.0, 10.0, 11.0, and 12.0V.

Create a spreadsheet table as shown below. You will have 13 rows of data.

	A	B	C	D	E
1	Volts at A	Volts at B	Volts Lamp	A Lamp	Watts Lamp
3	0	0	0	0	0
4	1				
5	2				

C4 has: =A4–B4   D4 has: =B4/10   Cell E4 has: =C4*D4

Analysis

1.   Use the spreadsheet to graph the current versus voltage characteristics of the LED. Current is on the vertical axis and voltage is on the horizontal axis. Label the graph.

2   Use the spreadsheet to graph the current versus voltage characteristics of the lamp Current is on the vertical axis and voltage is on the horizontal axis. Label the graph.

3.   The resistance of the device at a point on the graph is equal to the value of the voltage at that point divided by the value of the current at that point. Use the spreadsheet to calculate the resistance for each row in column F.

Plot the resistance of the LED as a function of power dissipation using your spreadsheet data (resistance on vertical axis).

Plot the resistance of the lamp as a function of power dissipation using your spreadsheet data (resistance on vertical axis). Why does the resistance of the lamp change as more power is dissipated by it?

4.   Copy each properly labeled graph and paste it into a word document. Your document should have the form below:

Your Name   Course Number   Date   Experiment Number and Name

Circuit diagrams with labeled voltages and currents
LED Current versus Voltage Graph
Lamp Current versus Voltage Graph
LED Resistance versus Power Graph
Lamp Resistance versus Power Graph

Brief summary (what you learned from this exercise).

The above is a short report. The instructor will specify how to turn in the report. The instructor may also ask for additional analysis, including a more comprehensive lab report.

Experiment 6: Analog "D'Arsonval" Meter

Introduction

An analog meter uses magnetic fields to measure electric current. A coil of wire wound on a bobbin creates a magnetic field whose strength is proportional to the wire's current. The bobbin is suspended in between the magnetic field of a permanent magnet as shown in the illustration on the right. The bobbin and attached needle rotate due to the magnetic force between the bobbin and the permanent magnet.

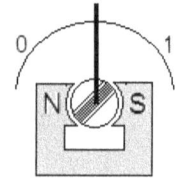

The wire winding has a resistance which determines one of the meter's characteristics. For example, if the winding resistance is 1000Ω, and one milliamp causes "full scale" deflection, its "sensitivity" will be given as "1000 ohms per volt". By Ohm's law, 1V applied to a 1000Ω resistor will create a current of one milliamp.

So a 1000Ω, one milliamp meter is also a voltmeter whose full scale voltage is 1V. The voltage range of the meter can be changed to 10V by adding a series resistance that will result in a one milliamp current when the applied voltage is 10V. The total resistance required to limit the current to one milliamp is $10,000\Omega$. Connecting a 9000Ω resistor in series with the meter makes the meter a voltmeter whose range is zero to ten volts.

Objectives

Measure the operating characteristics of an analog milliammeter. Use the measured meter characteristics to design a dual range analog voltmeter. Evaluate the performance of the voltmeter.

Procedure

> **Equipment and Parts**
>
> Power supply, DMM and Breadboard.
> Analog meter, 1mA < 1000Ω (or 100µA < $10,000\Omega$).
> R1: 1k, ¼ watt, 5% for 1mA meter (or 10k, ¼ watt, 5% for 100µA meter).
> Values of resistors R2 and R3 to be determined.

Part A: Meter Characteristics

1. Measure and record the value of R1. R1 is the 1k resistor if using a 1mA meter (or 10k if using a 100µA meter). Set the power supply voltage to zero. Connect the circuit on the right. Vs is a variable power supply (0 to 10V). Connect the DMM to read the voltage, Vm, across the meter.

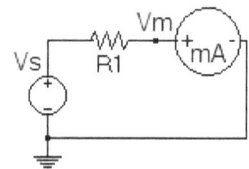

    R1: _____

2. Slowly increase the voltage, Vs, until the analog meter reads full scale. Your eye should be directly above the meter's needle to reduce "parallax error". Measure and record the voltage, Vm, across the meter.

Connect the DMM to Vs. Measure and record Vs.

Vm: _____   Vs: _____

3. Calculate the actual meter current, I_{ma}, and resistance, R_{ma}, using the voltages, Vm and Vs, and the measured value of R1.

$$I_{ma} = \frac{Vs - Vm}{R1} = \underline{\hspace{3cm}} \qquad\qquad R_{ma} = \frac{Vm}{I_{ma}} = \underline{\hspace{3cm}}$$

Part B: Meter Design

1. Use the calculated values of I_{ma} and R_{ma} from part A, step 3, to design a dual range voltmeter circuit, as shown on the right, to measure 1V full scale and 10V full scale. Calculate the required values R2 and R3.

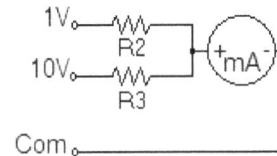

R2: _____   R3: _____

Use a series combination of resistors for R2 and R3 so that the resistors used in the circuit have measured values within 1% of the calculated values. Use a high value and low value resistor in series.

For example, if R2 = 930Ω, you could connect the standard values of 910Ω and 22Ω in series to obtain 932Ω. The percent error for this combination is:

$$\% \, Error = \frac{932 - 930}{930} \times 100\% = 0.215\%$$

Measure the resistance of your series combination. You may need to try another combination if your measured resistance is not within 1% of the required resistance. Note that 1% of 1000Ω is equal to 10Ω.

2. Test the voltmeter circuit by comparing the analog meter readings to the DMM readings as follows:

    a) Connect the DMM to the variable power supply. Connect the 1V input range of the analog meter (R2) to the variable power supply. The meters will be in parallel.

    b) Carefully set the power supply voltage so that the analog meter reads 0.20V. Observe the analog meter reading from directly above the needle to minimize parallax error.

    c) Read the voltage on the analog meter. Record result in the table on the next page.

    d) Repeat steps (b) and (c) for voltages of 0.40, 0.60, 0.80, and 1.00V.

    e) Repeat the procedure for the analog meter's 10V range and record results in the table on the next page (for 2.0, 4.0, 6.0, 8.0, and 10.0V).

1V RANGE			10V RANGE		
Analog Meter	DMM	% Error	Analog Meter	DMM	% Error
0.2			2.0		
0.4			4.0		
0.6			6.0		
0.8			8.0		
1.0			10.0		

Analysis

1. Calculate the sensitivity of the analog voltmeter in ohms per volt on the 1V and 10V range.

2. Calculate the input resistance of the analog voltmeter on the 1V and 10V range.

3. Calculate the percent error for all of the readings and record the results in the table. You could use a spreadsheet to do the calculations.

4. Calculate the average percent error of the analog meter on the 1V and on the 10V range.

5. Assuming that the analog meter is as accurate as the digital meter, calculate the voltage that each meter would read across the 15k resistor given that the digital meter has resistance of 1MΩ and the analog meter has a resistance of 10kΩ. Calculate the percent error for each meter.

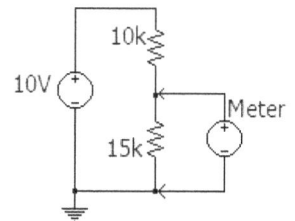

6. Use the values of I_{ma} and R_{ma} of the meter to design an analog ammeter with a full scale deflection current of 100mA. Calculate the required value of the shunt resistance (resistance in parallel with the meter). What is the net resistance of the 100mA meter (shunt resistance in parallel with the meter resistance, R_{ma}).

Experiment 7: DC Measurements and Meter Loading

Introduction

Meters used to measure voltage or current have an internal resistance. Since the meter must be connected to the circuit to make a measurement, the circuit is changed by the resistance of the meter. This is referred to as the meter's "loading effect".

Most meters have amplifiers built in so that only a small amount of power is needed from the circuit, and the loading effects are minimized. But this is not always the case.

A voltmeter is connected in parallel with the circuit across which the voltage is being measured. The loading effect of the meter will be minimal if the meter resistance is much larger than the circuit resistance. Ideally, the voltmeter resistance should be infinite. Practically, most electronic voltmeters (DMMs) have a resistance of 10 megohms.

The effect of the voltmeter resistance, Rm, across a circuit element, Rc, can be calculated using the parallel resistor equation:

$$Rt = \frac{Rc \cdot Rm}{Rc + Rm}$$

An ammeter is connected in series with the circuit in which the current is being measured. The loading effect of the meter will be minimal if the meter resistance is much smaller than the circuit resistance. Ideally, the ammeter resistance should be zero.

Practically, the resistance of most electronic ammeters (DMMs) varies with the range setting, with the highest ranges having the least resistance. This is one case where setting the range for the most significant digits may not always result in the most accurate reading.

The effect of the ammeter resistance, Rm, in series with circuit element, Rc, can be calculated using the series resistor equation:

$$Rt = Rc + Rm$$

Even if the meter loading effect is insignificant there will be an uncertainty in the measured value due to measurement errors caused by the accuracy of the meter.

Objectives

A series-parallel circuit will be connected on a solder-less breadboard. Voltage and current measurements will be made on the circuit and the results will be compared to theoretical expectations. The effect of the resistance of the meter will be determined.

Procedure

> ### Equipment and Parts
>
> DMM, Power Supply, and Breadboard.
> Resistors: 100, 220, 1 Meg, 2.2 Meg, ¼ watt, 5%.

Part A: Voltage Measurement

1. Measure and record the values of the resistors in the circuit. You will use these measured resistor values for your calculations and for LTspice input in experiment 8.

   R1: _____   R2: _____

   R3: _____   R4: _____

2. Connect the circuit below on the left and set the power supply voltage to 6.0V.

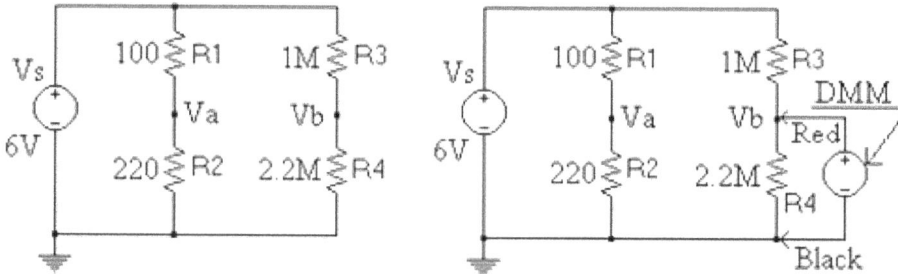

. Lay out the circuit on the breadboard so it looks similar to the schematic diagram. Use the minimum number of wires to connect the circuit.

   Note that the positive power supply lead is connected to the top of R1 and the that negative power supply lead is connected to the bottom of R2.

   The voltmeter is connected across R4.

3. Measure and record Va, Vb and Vab (Vab = Va – Vb).

   Va: _____   Vb: _____   Vab: _____

4. Obtain and record the internal resistance, Ri, of your voltmeter. Ri: _____

Part B: Current Measurement

1. The circuit on the right shows the DMM connected in series with the 100Ω and 220Ω resistors.

   You need to break the connection between R1 and R2 and insert the meter as shown. The meter will read the current through the 100Ω and 220Ω resistors.

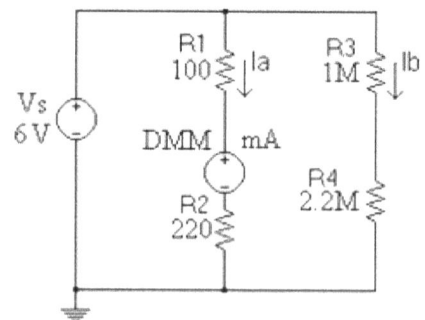

2. The internal resistance of a DMM on the current ranges varies with the range. Check the manual on your DMM to see what its internal resistances are on the current ranges. If you don't have a manual, your instructor should provide the information. Record the meter resistances below.

Rm(0.2mA): _____   Rm(2mA): _____

Rm(20mA): _____   Rm(200mA): _____

3. Measure and record the current through the 100Ω and 220Ω resistors with the meter on the 20mA range.

Ia_{20}: _____

4. Measure and record the current through the 100Ω and 220Ω resistors with the meter on the 200mA range.

Ia_{200}: _____

Analysis, Part A

1. Calculate the theoretical values of the voltages: Va, Vb, and Vab, without taking the meter resistance into account. Calculate the percent error between the theoretical and measured results.

2. Calculate the voltages: Va, Vb, and Vab taking the meter resistance into account. Calculate the percent error between the measured and calculated results.

Analysis, Part B

1. Calculate the theoretical current, Ia, (without taking the meter resistance into account). Calculate the percent error between the theoretical and measured values (both current ranges).

2. Calculate the current, Ia, taking the meter resistance into account. Calculate the percent error between the calculated and measured results.

3. Briefly explain the significance of the error analysis in steps 1 and 2 above.

Experiment 8: LTspice Circuit Simulation

Introduction

LTspice IV will be used to simulate the circuit of experiment 7. It is important to always use your measured part values so that you can compare your measured results to your simulation results.

Drawing the Circuit

1. Start the *LTspice* program. In the main menu bar, click on File and select New Schematic.

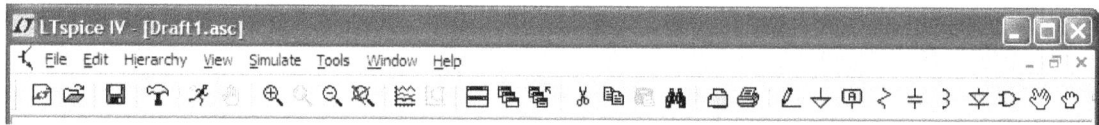

2. Create the new schematic. Left click on the resistor symbol and drag and place the resistor, R1. Get three more resistors and place them as shown on the right.

    Right click on the "*R*" of each resistor. Change the values to the measured values from Experiment 7.

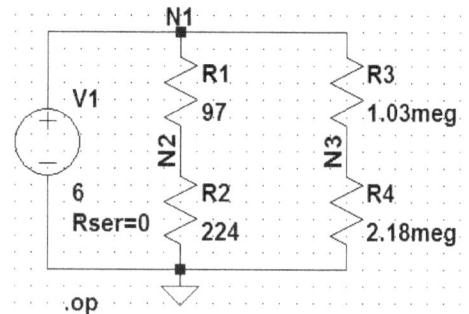

3. Left click on the gate symbol between the diode and the hand in the main menu to get the dialog box on the right. Select the part "*voltage*". place it in the schematic.

    Click on the ground symbol and place it under R2.

4. Left click on the pencil in the main menu to connect the components.

5. Left click on the *A* between the ground symbol and the resistor symbol to label the nodes *N1, N2,* and *N3* as shown on the schematic by dropping the dot on the nodes.

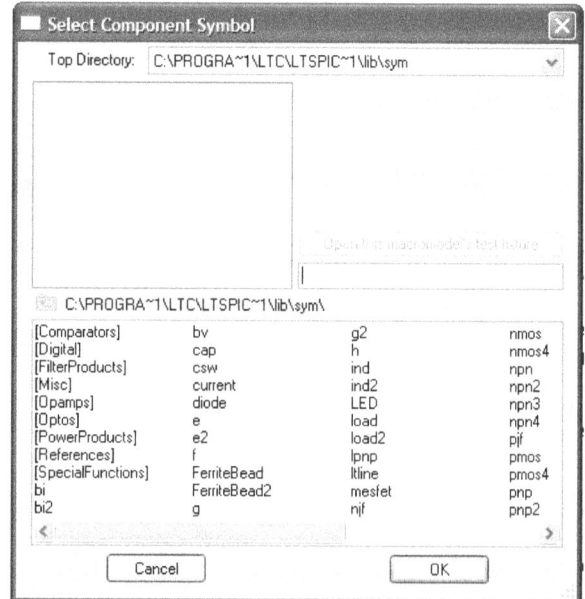

Click "Help" in the menu for more information on creating the schematic.

6. Click on Simulate and then select the *Edit Simulation Cmd* in the menu. Select *DC op pnt asshown on the right.* Click on *ok*. A ".op" command will appear which can be placed anywhere on the schematic.

7. Click on *Simulate* and then select *Run*. If there are no errors, you will see an "*Operating Point*" file.

   The Spice netlist shows the connections of the parts, part models, and types of analysis to be performed.

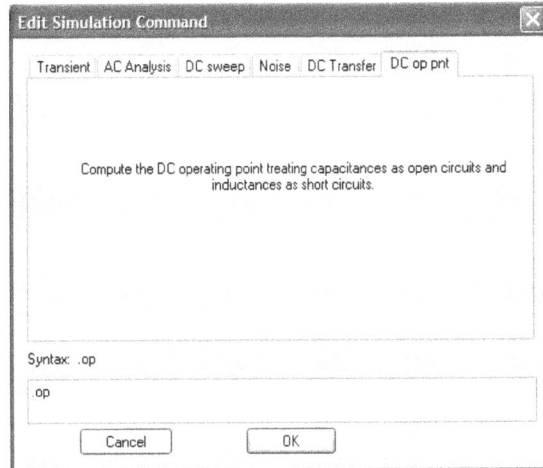

Edit Simulation Command

Transient | AC Analysis | DC sweep | Noise | DC Transfer | DC op pnt

Compute the DC operating point treating capacitances as open circuits and inductances as short circuits.

Syntax: .op

.op

Cancel | OK

--- Operating Point ---

V(n1):	6	voltage
V(n2):	4.18692	voltage
V(n3):	4.07477	voltage
I(R4):	1.86916e-006	device_current
I(R3):	1.86916e-006	device_current
I(R2):	0.0186916	device_current
I(R1):	0.0186916	device_current

Netlist

```
R1 N1 N2 97
R2 N2 0 224
R3 N1 N3 1.03meg
R4 N3 0 2.18meg
V1 N1 0 6 Rser=0
.op
.backanno
.end
```

Exercise and Analysis

1. Simulate the loading effect of the voltmeter at nodes 2 and 3 using a resistor whose resistance is equal to the meter's internal resistance. Compare simulated results to Experiment 7 measured results.

2. Simulate the loading effect of the ammeter using a resistor whose resistance is equal to the meter's internal resistance. Compare simulated results to Experiment 7 measured results.

Experiment 9: Kirchhoff's Voltage and Current Laws

Introduction

Kirchhoff's voltage and current laws are very important in circuit analysis. Kirchhoff's voltage law says that what goes up must come down. If you take any path around a circuit and return to where you started, all of the voltage drops in that path should equal all of the voltage rises. The algebraic sum of the voltages for any closed path must equal zero.

Kirchhoff's current law says that what goes in must come out. All of the currents entering a node must equal all of the currents leaving a node.

$$\text{KVL:} \sum_{i=1}^{n} V_i = 0 \qquad \text{KCL:} \sum_{i=1}^{n} I_i = 0$$

Objectives

In previous lab exercises we have made use of Kirchhoff's voltage law when we summed voltages in series circuits. This exercise will focus on using Kirchhoff's voltage law and Kirchhoff's current law in parallel and series-parallel circuits.

Procedure

Equipment and Parts
Power Supply, DMM, and Breadboard Resistors: 100, 220, 330, and 470Ω, all ¼ watt, 5%

Part A: Parallel Circuits

1. Measure and record the values of the resistors that you will use in the circuit.

   R1_____   R2_____   R3_____

2. Connect the circuit on the right so that it will be easy to insert the DMM in series with each component for measuring current. This circuit has three parallel branches. The voltage across each branch is exactly 4.0V.

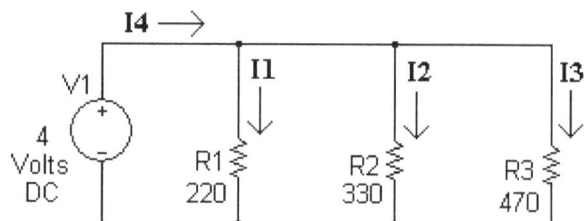

Be very careful when switching the meter between measuring voltage and measuring current. Trying to measure voltage while in current mode may cause damage to the meter, circuit, or the power supply.

DMMs usually have a fuse in their current measuring circuit. If your meter does not read current, it may be that the fuse needs replacement.

3. Carefully measure and record the currents, I1, I2, I3, and I4. Use the 20mA meter range. Note and record the meter's resistance, R_m, on the 20mA range.

   R_m: _____ (Refer to the instrument manual or ask the instructor.)

   I1:_____   I2: _____   I3: _____   I4: _____

4. Prepare a spreadsheet for data entry as shown below. R4 will be used in part B of this exercise. Rows 11 through 15 of this spreadsheet will be used for part B.

	A	B	C	D	E	F
1		R1	R2	R3	R4	
3	Measured R					
5	PART 1	I1	I2	I3	I4	
6	Measured I					
7	Calculated I					
8	% Error					

Part B: Series-Parallel Circuits

1. Use the same resistors for R1, R2, and R3 as you used in part A. Measure and record the value of R4, the 100Ω resistor.

   R4_____

   Connect the circuit on the right. Set the source voltage to exactly 10.0 volts.

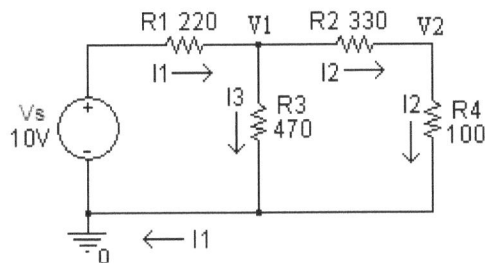

2. Measure and record the voltages, V1 and V2.

   V1_____   V2_____

3. Currents: I1, I2, and I3 will be determined by using the voltage drop across the resistors and Ohm's law.

$$I1 = \frac{10 - V1}{R1} \qquad I2 = \frac{V1 - V2}{R2} = \frac{V2}{R4} \qquad I3 = \frac{V1}{R3}$$

These equations may be entered into the spreadsheet cells E14, F14, and G14, using the appropriate cell references for the voltages and resistances. Refer to the analysis section for part B.

Analysis, Part A

1. Have the spreadsheet calculate the currents, I1, I2, and I3. Be sure to use the measured values of your resistors in the spreadsheet row 3. Also have the spreadsheet calculate the percent error in the measurements.

For example, enter the equation: =4/B3 into cell B7, and: =100*(B6-B7)/B7 into cell B8.

2. Explain the most probable cause of the errors.

3. Show that Kirchhoff's current law holds (I4 = I1 + I2 + I3) exactly for the calculated currents, and approximately for the measured currents.

Analysis, Part B

1. Calculate the voltages V1, and V2; record the results in the spreadsheet.
    Verify the calculations with an *OrCAD PSpice* simulation. Have the spreadsheet calculate the percent error for each measurement.

	A	B	C	D	E	F	G
11	PART 2	V1	V2		I1	I2	I3
13	Measured				X	X	X
14	Calculated						
15	% Error				X	X	X

2. Have the spreadsheet calculate the currents I1, I2, and I3 from the measured voltages, V1 and V2.

3. Show that the currents satisfy Kirchhoff's current law.

4. Explain why the measured results of part B are more accurate than the measured results of part A.

5, Show that the voltages across R1, R2, and R4 add up to 10V and satisfy Kirchhoff's voltage law.

History Note:

Gustav Robert Kirchhoff (12 March 1824 – 17 October 1887) was a German physicist who contributed to the fundamental understanding of electrical circuits, spectroscopy, and the emission of black-body radiation by heated objects. He coined the term "black body" radiation in 1862, and two sets of independent concepts in both circuit theory and thermal emission are named "Kirchhoff's laws" after him, as well as a law of thermochemistry.

From Wikipedia, the free encyclopedia

Experiment 10: Potentiometer Voltage Dividers

Introduction

Potentiometers (pots) are very commonly used as variable voltage dividers. A potentiometer is basically a resistor with a sliding contact so that the resistance between the sliding contact and the ends of the potentiometer can be varied. Most pots are the single turn rotary type. Multi-turn rotary types and "slider" types are also available.

The schematic diagram of the potentiometer is shown below on the left in figure a. Figure b shows how the rotary type operates, while figure c shows how the slider type operates.

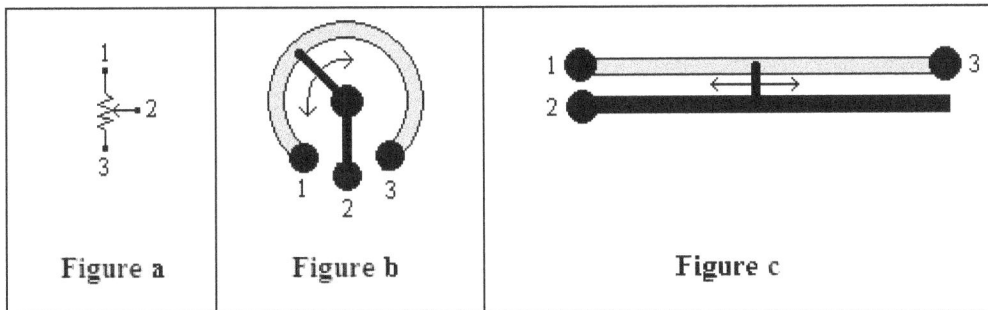

| Figure a | Figure b | Figure c |

Potentiometers are available in a wide range of resistance and power capabilities. The resistance rating is the total resistance (also the maximum resistance) of the potentiometer between terminals 1 and 3 as shown above. The slider moves between terminals 1 and 3. As the slider moves from terminal 1 toward terminal 3 the resistance between terminals 1 and 2 increases and between terminals 2 and 3 decreases.

The resistance change may be proportional to the distance moved (linear taper pot) or may change logarithmically with the distance moved (audio/log taper pot).

Potentiometers may have a rotary shaft with a knob on it or they may have a screwdriver slot for screwdriver adjustment. The screwdriver types are usually called "trimmer pots" and are often used to calibrate "trim" circuits. The type used in this lab is a low power (1/4 or ½ watt) trimmer type.

Objectives

The purpose of this exercise is to become familiar with the potentiometer as a variable voltage divider. In addition, the voltage divider loading effect will be investigated.

Procedure

Equipment and Parts

Power Supply, DMM, and Breadboard.
Resistors: 4.7K, and two 1K, ¼ watt, 5%.
Trimmer Potentiometers: 100Ω , 1kΩ, and 10kΩ, ¼ or ½ watt.

Part A: Basic Measurements

1. Measure the values of R4 and R5.

    R4_____

    R5_____

2. Neatly connect the circuit on the right. Include the "jumper" wires designated as JP1 through JP4 so that it will be easy to disconnect each potentiometer and load resistor.

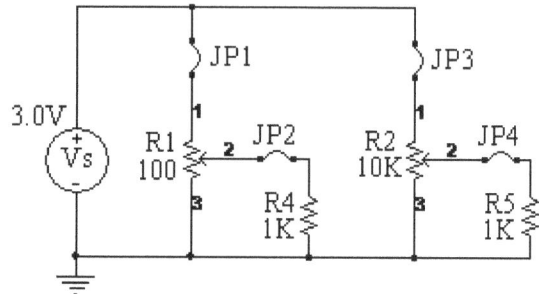

3. With jumpers JP1 and JP3 in place, remove jumpers JP2 and JP4 so that the 1KΩ load resistors are disconnected from the pots. Set the power supply voltage to exactly 3.00V. Adjust each pot so that the voltage at each terminal 2 is exactly 1.50V.

4. Remove jumpers JP1 and JP3 so that the pots are disconnected from the power supply. Be careful to not disturb the pot settings. Measure and record the resistance of each pot from terminal 1 to ground, and from each terminal 2 to ground.

    $R1_{13}$ is the resistance of R1 between terminals 1 and 3, and $R1_{23}$ is the resistance of R1 between terminals 2 and 3

.   Similarly, $R2_{13}$ is the resistance of R2 between terminals 1 and 3, and $R2_{23}$ is the resistance of R2 between terminals 2 and 3.

    $R1_{13}$ _____ $R1_{23}$ _____

    $R2_{13}$ _____ $R2_{23}$ _____

5. Being careful not to disturb the pot settings, replace jumpers JP1and JP3 so that the pots are connected to the power supply. Verify that the voltage on terminal 2 of each pot is still 1.50V.

6. Being careful not to disturb the pot settings, replace jumpers JP2 and JP4 so that terminal 2 of each pot is connected to a 1KΩ load resistor. Make sure that the power supply voltage is still 3.00V. Measure and record the voltage at terminal 2 of each pot with respect to ground. $VR1_{23}$ is the voltage at terminal 2 of R1 to ground and $VR2_{23}$ is the voltage at terminal 2 of R2 to ground.

    $VR1_{23}$ _____ $VR2_{23}$ _____

Part B: Changing the Pot's Range of Adjustment

Suppose that it is desired to get a voltage variable from 1V to 2V by turning the pot from one end to the other using the 3.00V power supply. Of course that voltage range is available from the circuit of part A, but a 1V change would correspond to rotating the pot over just one third of its range. We will investigate a circuit in this part of the exercise that will produce a 1V to 2V change over the pot's entire range.

1. Use the measured values of the resistors, R1, R4 and R5 to be used in the circuit on the right. Measure the value of R7.

   R7_____

   Connect the circuit on the right. Set Vs to exactly 3.00V.

2. Disconnect the jumper, JP1. Measure and record the minimum and maximum value of the voltage Vo by turning the pot to both ends of its range.

   Vo(min)_____   Vo(max)_____

3. Connect the jumper, JP1, and measure and record the minimum and maximum value of the voltage Vo by turning the pot to both ends of its range.

   Vo(min)_____   Vo(max)_____

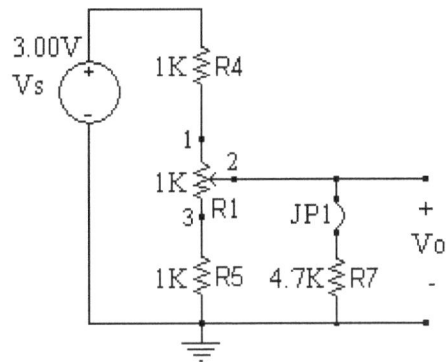

Analysis, Part A

Be sure to use the measured values of the resistors to do the theoretical calculations.

1. Use Ohm's law (voltage divider equation) and the measurements in part A, step 4, to calculate the voltages $VR1_{23}$ and $VR2_{23}$.

   Comment on the accuracy of the measurements and the loading effect of the voltmeter.

2. Calculate the voltages $VR1_{23}$ and $VR2_{23}$ with the 1K resistors connected. Compare your results to the measurements in part A, step 6. Calculate the percent error caused by the loading effect of the 1K resistors on the measurements.

   Which measurement had the greatest loading effect and why?

Analysis, Part B

Be sure to use the measured values of the resistors to do the theoretical Calculations.

1. Calculate the theoretical values of Vo(min) and Vo(max) in part B, step 2. Calculate the percent accuracy of these measurements.

2. Calculate the theoretical values of Vo(min) and Vo(max) in part B, step 3. Calculate the percent accuracy of these measurements.

Optional challenging problem:

Note the effect of the 4.7K load resistance and redesign the circuit by changing the values of the fixed resistors in the voltage divider so that the voltage range is about 1V to 2V with the 4.7K load resistor connected.

Hint: Note that Vo1 must be 1V in figure A and Vo2 must be 2V in figure B. You will need two simultaneous equations solve for R1 and R2.

Use simulation to verify the results of step 3 above.

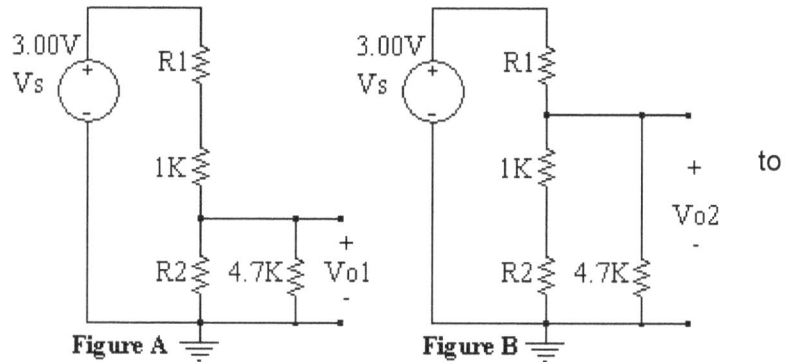

Figure A

Figure B

LTspice Simulation Example

A suggested circuit for analyzing this lab exercise is given below. The values of R1 and R4 should be the same, and the values of R3 and R6 should be the same. The values of R2 and R5 should be the measured value of the pot resistance, and R7 and R8 should have the measured value of the 4.7kΩ resistor.

Try the calculated values for R1 and R3 below. Note that R1 = R4 and R3 = R6. Compare the resulting voltages at nodes N1 and N2 to the values you calculated.

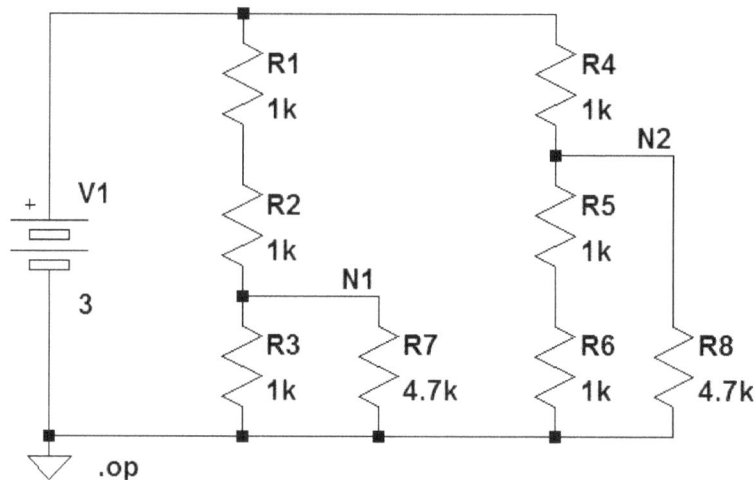

The operating point results below are for the resistor values given in the circuit above.

```
        --- Operating Point ---
    V(n001):        3               voltage
    V(n002):        1.93789         voltage
    V(n1):          0.875776        voltage
    V(n2):          1.75155         voltage
    V(n003):        0.875776        voltage
```

Experiment 11: Thevenin's Theorem and the Bridge Circuit

Introduction

Wheatstone bridge circuits are used in many applications in electronics. A common problem is to find the voltage across the center element (R5 on the right) and the current through it. R5 could represent a load, or a device drawing power from the circuit. Without yet having knowledge of the node voltage method, one would not be able to solve for R5's voltage or current. Fortunately, Thevenin's theorem can be used instead and it is a much simpler method, especially when solving for many different values of R5.

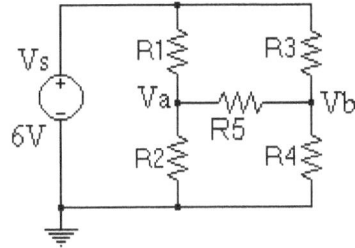

The Thevenin voltage of the circuit above "external" to R5 is found by determining the "open circuit voltage", Va – Vb, with R5 removed.

When R5 is removed, Va and Vb can be calculated with the voltage divider equation.

$$Va = \frac{R2}{R1+R2} Vs \quad \text{and} \quad Vb = \frac{R4}{R3+R4} Vs$$

The Thevenin voltage can be calculated as V_{Th} = Va – Vb.

To find the Thevenin resistance of the circuit from the perspective of R5, R5 is removed and the voltage source is set to zero (replaced by a short circuit). Once that is done, the circuit drawing can be rearranged for clarity. Refer to the diagrams below.

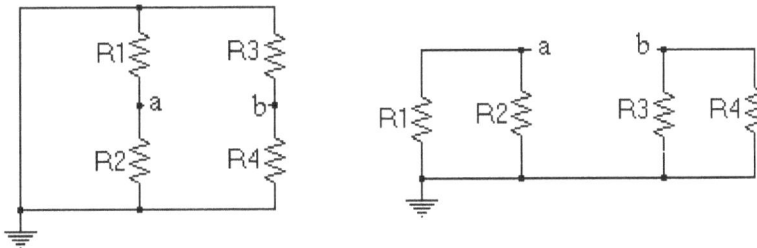

It can be shown that the Thevenin resistance is:

$$R_{Th} = (R1\|R2) + (R3\|R4) = \frac{R1 \cdot R2}{R1+R2} + \frac{R3 \cdot R4}{R3+R4}$$

According to Thevenin's Theorem, the Thevenin equivalent circuit on the right will produce the same voltage across R5, for any value of R5, as the original bridge circuit.

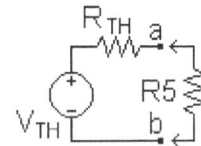

Objectives

A comparison will be made between the solutions for the voltage across the resistor, R5, using the node voltage method and the Thevenin equivalent circuit method. The Thevenin equivalent circuit will be determined by measurement as well as by calculation.

The value of the center element, R5, will be varied in both circuits to show that they are equivalent. The Thevenin equivalent circuit will also demonstrate the maximum power transfer theorem.

Procedure

Equipment and Parts
Power Supply, DMM, and Breadboard. R1 = 1500Ω   R2 = 2200Ω   R3 = 1000Ω   R4 = 470Ω R5 values: 390, 680, 1200, 2200, and 3900Ω.

1. Measure and record the values of all the resistors you will be using.

   R1 _____ $R5_{390}$ _____

   R2 _____ $R5_{680}$ _____

   R3 _____ $R5_{1200}$ _____

   R4 _____ $R5_{2200}$ _____

                                 $R5_{3900}$ _____

2. Set the DMM to read DC volts. Use the meter range which gives you at least three significant digits. Connect the circuit in figure A below. Use R5 = 390Ω initially.

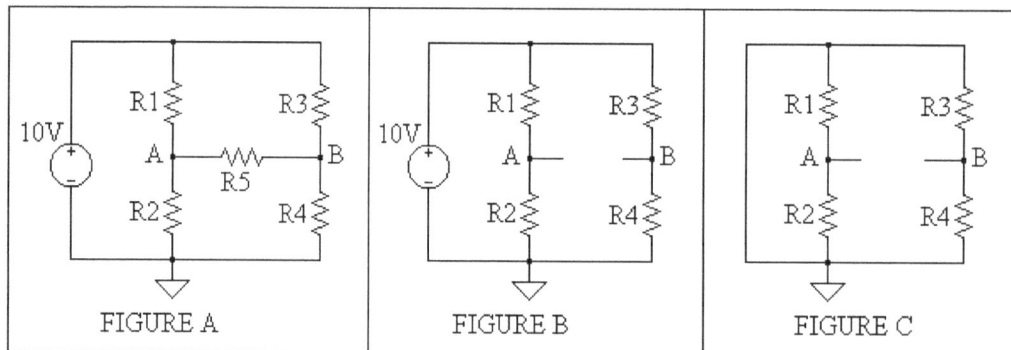

FIGURE A   FIGURE B   FIGURE C

3. Set the power supply voltage to 10.0V.

4. Measure and record the values of the voltage V_{AB} in the table on the right for the given values of R5.

   Since V_{AB} is equal to $V_A - V_B$, it can be measured with the positive probe on V_A and the negative probe on V_B.

R5 (labeled)	Vab
390Ω	
680Ω	
1200Ω	
2200Ω	
3900Ω	
∞ Ω (open)	

5.   Turn off the power supply and disconnect it from the circuit. Then connect a wire from the top of R1 to the bottom of R2, as shown in figure C.

    Use the ohmmeter to measure the resistance of the circuit between nodes A and B. This resistance is the measured Thevenin resistance of the circuit for nodes A and B.

6.   Record the Thevenin resistance, R_{TH}, from step 5 above. Record the voltage Vab with $R5 = \infty$ from step 4 as the Thevenin voltage, V_{TH}.

7. R_{TH}: _____ V_{TH}: _____

8.   Use the values in step 7 above to connect a Thevenin equivalent of the bridge circuit for nodes A and B. Use a combination of resistors for R_{TH} to match the resistance recorded in step 7 to within one percent.

    The exact value may be obtained by connecting your bridge circuit resistors as shown below.

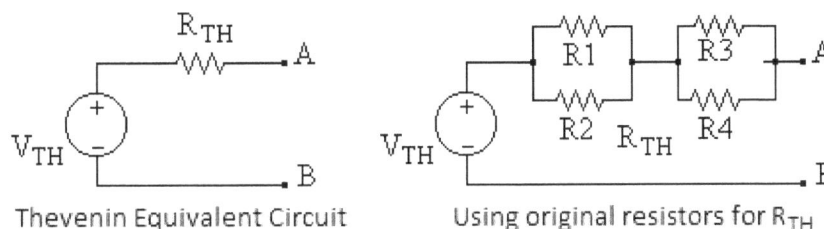

Thevenin Equivalent Circuit   Using original resistors for R_{TH}

9.   Set the power supply voltage to V_{TH} from part 6. Connect the various R5 resistors between points A and B in the Thevenin equivalent circuit.

    Measure and record the voltage V_{AB} in the table on the right.

R5	Vab
390Ω	
680Ω	
1200Ω	
2200Ω	
3900Ω	

Analysis

1.   Enter the data for the voltage V_{AB} into a spreadsheet as shown below:

	A	B	C	D	E
1	R5 (measured)	Vab (Step 4)	Vab (Step 9)	% Error	Power (Step 9)
2	390				
3	680				
4	1200				
5	2200				
6	3900				
7	∞ Open				

2. Use the measured values of R5 in column A. Use the equation for percent error in cell D2, and use "fill down" to have the spreadsheet calculate the error for cells D2 to D7.

Similarly, enter the equation for power in cell E2 and use "fill down".

% Error = ((2 - B2) / B2) *100   Power = C2*C2 / A2

3. Solve the circuit in figure A for the voltage V_{AB}, for R5 equal to 1200 Ω, using the node voltage method.

4. Simulate the circuit of figure A. Generate a list of values of Vab for all 5 of your measured values for R5. Use the spreadsheet to calculate the percent error between the simulated and measured results (measured values from step 5).

LTspice Simulation of a Bridge Circuit and Variable Resistor

In the circuit simulation below, the value of R5 was varied from 100Ω to 4000Ω in 100Ω steps. R5's value is indicated as "{RX}" on the schematic.

Get the Parts

Get and place the resistors. Get and place the voltage source and ground. Connect the circuit. Label nodes N1, N2, and N3.

Setup the Analysis

Spice directives can be entered directly into the schematic by pressing the letter "s". Enter the directives ".op" and ".step param RX 100 4000 100" and place them in a convenient place on the schematic. Value of R5 is varied from 100Ω to 4000Ω in 100Ω steps.

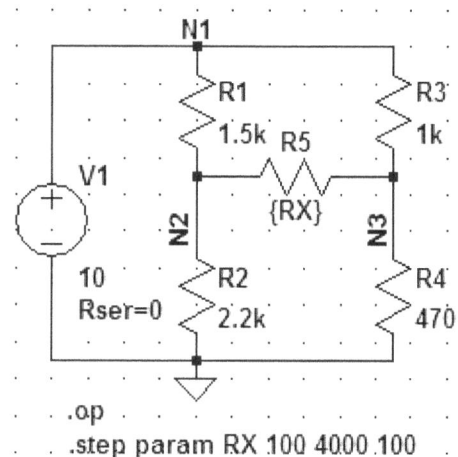

Simulate the Circuit

Click *Run* under the *Simulate* menu. Move the mouse to N2, left click and drag the probe to N3. This will display the voltage across R5. To plot of the power dissipation of R5, click in the plot window, click on *Plot Settings* in the main menu, select *Add trace*, type V(n2,n3)*-I(R5), click *ok*.

Plot on the right shows that maximum power occurs for a value of R5 of about 1200 Ω.

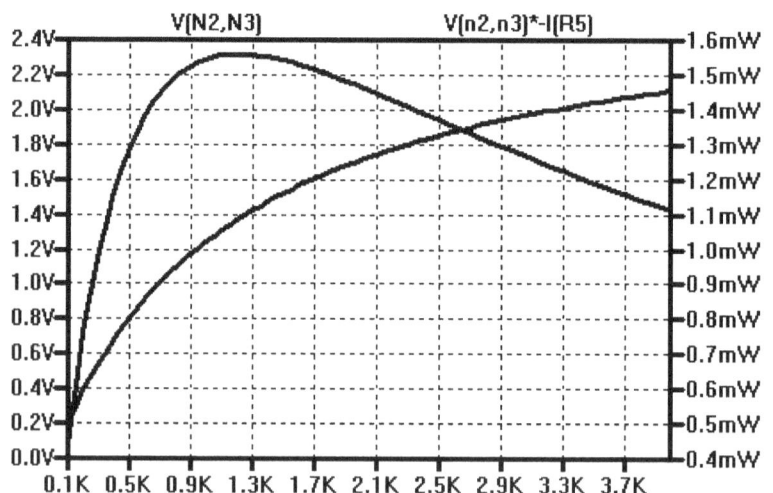

38

The values of the voltage across R5 may obtained using the mouse in LTspice.

Select the graph window.   Move the "+" along the graph to each desired value of R5 and note the voltage across R5 and its power dissipation for each value of R5. The value appears at the bottom of the LTspice window.

When the + is moved to 1.2k on the power trace, the display at the bottom of the widow reads: x = 1.198K   y = 2.315V, 1.557mW.   This indicates that the power is 1.557mW, but the voltage is not 2.315V.

Similarly, when the + is moved to 1.2k on the voltage trace, the display at the bottom of the widow reads: x = 1.198K,   y = 1.366V, 1.083mW.   This indicates that the voltage is 1.366V, but the power is not 1.083mW.

The voltage and power traces could have been displayed in separate windows to avoid the confusion.

History Note:

Léon Charles Thévenin (30 March 1857 - 21 September 1926, Paris) was a French telegraph engineer who extended Ohm's law to the analysis of complex electrical circuits.

From Wikipedia, the free encyclopedia

Experiment 12: Transistor / Dependent Current Source

Introduction

One of the greatest inventions of the 20th century is the transistor. It has allowed the miniaturization of electronic circuits by making possible the integrated circuit or IC. It has revolutionized computers, and prompted many other technological advances.

In this experiment you will see how the transistor acts as a current controlled current source and how it can be modeled that way. The transistor used in this lab has three terminals that are called the emitter, collector, and base. The property of the transistor that makes it so important is that a small current flowing into the base can control a much larger current flow between the emitter and collector.

The relationship between the base current and the emitter to collector current is called the transistor's "current amplification factor", beta (β). β is typically between 100 and 300.

Schematic symbol, circuit model, and PSpice model of the transistor are shown on the right.

$Ic = \beta \cdot Ib$

$Ie = Ic + Ib$

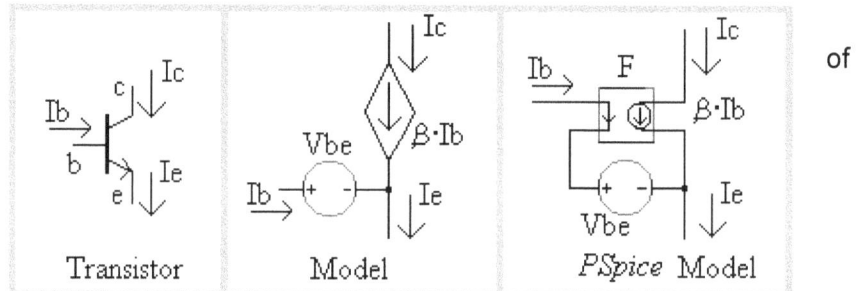

Ib is the base current, Ic the collector current, and Ie is the emitter current. Note that the equation, Ie = Ic + Ib, is an expression of Kirchhoff's current law.

Voltage source, Vbe, is added to the models to simulate the voltage between the transistor's base and emitter, which is typically between 0.65V and 0.75V.

Objectives

A simple transistor circuit will be evaluated as a current controlled current source and as a constant current source. The transistor's current gain, β, will be measured. This parameter will be used in the dependent current source model in *PSpice*. The current control ability of the transistor will be measured and compared to *PSpice* or LTspice simulation.

Procedure

Equipment and Parts
DMM, Power Supply (15V and 0 to 6V), and Breadboard, Transistor: 2N3904 or 2N2222. Resistors: 100, 470, 1K, 2.2K, 4.7K, 6.8K, 8.2K, 10K, 22K, 100K.

Part A: Dependent Current Source

Measure the values of your resistors.

Labeled Value	Measured Value	Labeled Value	Measured Value
100		6.8K	
470		8.2K	
1K		10K	
2.2K		22K	
4.7K		100K	

1. Connect circuit on the right. Q1 is the transistor. Note that Q1's emitter lead is on the left when the transistor's flat side is facing you. Set V1 to 0V and V2 to 15.0V.

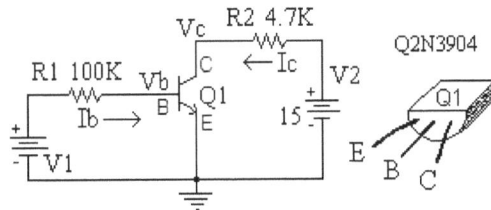

2. Connect the DMM to measure Vc. Adjust V1 to get the following voltages for Vc: 1, 2, 3, 4, 5, 6, 7, 8, 9, 10, 11, 12, 13, and 14. For each value of Vc, measure and record V1 and Vb. Set up a spreadsheet table as shown on the right to record the voltages.

3. Copy your results into a spreadsheet and have the spreadsheet calculate Ib, Ic, and β for each value of Vc. See format on the right..

Ib = (V1 - Vb) /100000)   Ic = (15 – Vc) / (4700)   β = Ic / Ib.

Part B: Constant Current Source

1. Make sure V2 is still exactly 15.0V and adjust V1 so that Vc is exactly 7.0V, as shown on the right. Change R2 to the following values, each time record Vc: 100, 470, 1K, 2.2K, 6.8K, 8.2K, 10K, 22K.

2. Copy your results into a spreadsheet and have the spreadsheet calculate Ic for each value of Vc. See format on the right.   Ic = (15 – Vc) / R2.

Analysis, Part A

1. Use the spreadsheet of part A to plot the collector current, Ic, of the transistor as a function of its base current, Ib (Ic on the vertical axis).

2. Simulate the circuit for part A. Use a current controlled current source to show that the transistor acts as a current controlled current source by plotting the current of the current source as a function of its controlling current.

    Set the gain of the current source to the average value of β (obtained by averaging the β results for Vc values of 4, 5, 6, 7, 8, 9, and 10 volts). See example below.

3. Compare your spreadsheet plot to your simulation plot. Over what range of Ic do the plots approximately match? Over what range of Ic do they not match?

Analysis, Part B

1. Use the spreadsheet of part B to plot the collector current, Ic, of the transistor as a function of R2 (with Ic on the vertical axis).

2. Simulate the circuit for part B. Refer to the example simulation on the next page. Use a current controlled current source for the transistor to show that it does act as a constant current source.

    Compare measured results to *LTspice* results. Where does the transistor fail as a constant current source?

    a) Use your measured resistor values and average value of β (from step 2 of the analysis of part A) for the gain of the current controlled current source.

    b) Use "DC sweep" to obtain value of Ic for each value of R2.

Example LTspice simulation of Part A

Connect the circuit as shown on the right. Current controlled current source, F1, is given the value "*V1 120*". This makes F1's current equal to the current through V1 times 120.

Set the simulation to "DC Sweep" and to sweep V2 from 0 to 6V in 0.1V steps.

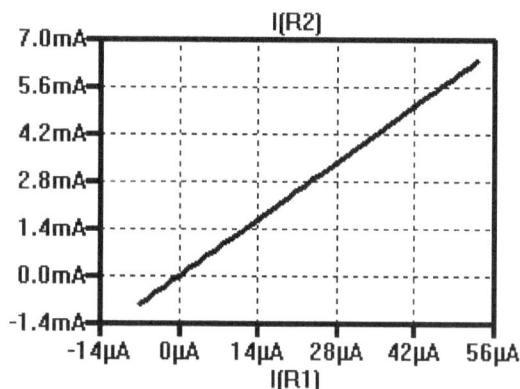

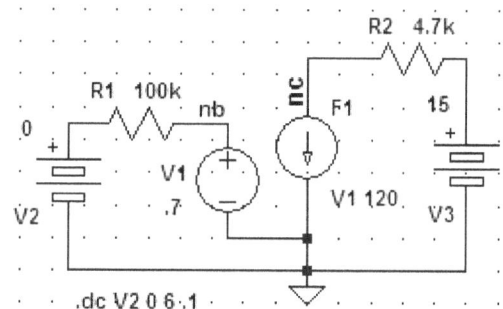

The vertical axis of the plot is the current through R2 which represents the transistor's collector current, Ic.

The horizontal axis of the plot is the current through R1 which represents the transistor's base current, Ib, in this model

Set the gain of the current source, F1, to that of your transistor to compare your measured results to the simulation.

To generate the plot on the previous page, the horizontal axis variable was changed from V(V2) to I(R1) by clicking just below the horizontal axis. This opens the horizontal axis dialog box .

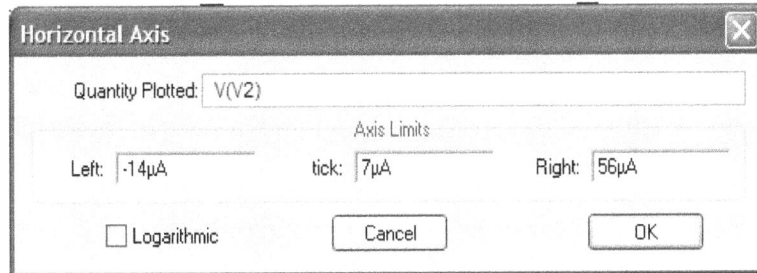

Horizontal Axis

Quantity Plotted: V(V2)

Axis Limits

Left: -14µA   tick: 7µA   Right: 56µA

☐ Logarithmic   Cancel   OK

Change the "Quantity Plotted" to I(R1). This generates a plot of I(R2) as a function of the current through R1.

History Note:

Thomas Alva Edison (February 11, 1847 – October 18, 1931). American inventor, scientist, and businessman. Developed the phonograph, the motion picture camera, and a practical electric light bulb.

From Wikipedia, the free encyclopedia

Experiment 13: Mesh Current Analysis

Introduction

Mesh current analysis is based on Kirchhoff's voltage law: the algebraic sum of the voltages around any closed loop or path must be zero. The example circuit below has three meshes (closed loops).

$$V1 - I1\,R1 - (I1 - I2)\,R2 = 0$$
$$-(I2 - I1)\,R2 - I2\,R3 - (I2 - I3)\,R4 = 0$$
$$-(I3 - I2)\,R4 - I3\,R5 + V2 = 0$$

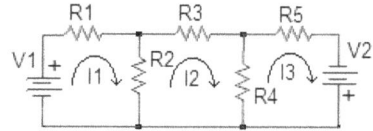

The application of Kirchhoff's voltage law results in the three independent simultaneous equations shown above. Refer to your textbook or instructor for more details, if needed.

The example above uses the three smallest meshes for the currents I1, I2, and I3. However, consider the circuit with a dependent current source shown below. The voltage across the current source is unknown. Kirchhoff's voltage law is applied to a larger loop, by-passing the current source. This is called the "super mesh". A second equation is obtained using the current source's dependency equation, as shown below.

$$V1 - I1\,R1 - I2\,R3 - I2\,R4 = 0 \quad \text{and given}: \quad Ix = 0.5 \cdot I2$$

$$\text{Apply Kirchhoff's Current Law}: \quad I1 - Ix - I2 = 0 \implies I1 = 1.5 \cdot I2$$

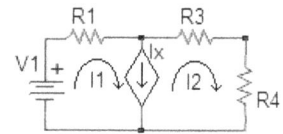

In this lab, a transistor will be used as our dependent source. Knowledge of how a transistor works is not required.

Objectives

The primary objective of this lab experiment is to gain experience in the application of mesh current analysis. Measurements will be made with the DMM and the results will be compared to theoretical calculations. In part A mesh analysis will be applied to a three resistor mesh network. In part B super mesh analysis will be applied to a circuit with a dependent source.

Procedure

Equipment and Parts
DMM, Power supply, and breadboard Resistors: 470, 750, 1.2k, two 1.5k, all ¼ watt, 5%. Transistor: 2N3904

Part A: Mesh Analysis – Resistor Network

1. Measure the values of the resistors.

    R1$_{750}$ _____   R2$_{1.5k}$ _____   R3$_{470}$ _____

    R4$_{1.5k}$ _____   R5$_{1.2k}$ _____

2. Connect the circuit on the right. Layout the circuit in the same form as the diagram. Do not use wires between the resistors.

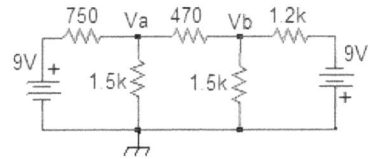

3. Set the plus and minus power supply voltages to exactly 9.0V.

4. Measure and record the voltages Va and Vb.

Va _____ Vb _____

Part B: Super Mesh Analysis – Dependent Current Source

1. Connect the circuit on the right. The transistor is a dependent current source. Its collector to emitter current is controlled by its base current, Ib.

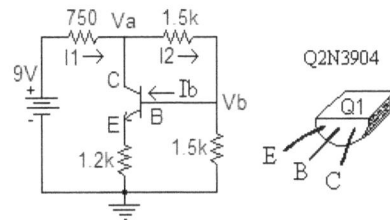

2. Measure and record the voltages Va and Vb.

Va _____ Vb _____

3. Measure and record the voltage at the emitter, V_E.

V_E _____

Analysis, Part A

1. Use mesh analysis to solve the circuit in part A using your measured part values. Determine the net current in each of the 1.5k resistors and solve for the voltages Va and Vb.

2. Calculate the percent difference between the measured and calculated values.

Analysis, Part B

1. The transistor and 1.2k resistor in the circuit can be replaced by a dependent current source whose current depends on the voltage Vb. Given that $V_E \cong Vb - 0.7$, write an equation for the current of the current source (Ix). The current Ib is negligible in this circuit and can be ignored.

2. The resulting circuit should look like the one on the right. Replace the 0.7V in the equation for Ix in step 1 above with your measured value of V_{bE} ($V_{bE} = Vb - V_E$). Use your measured resistor values to calculate an accurate value of Ix.

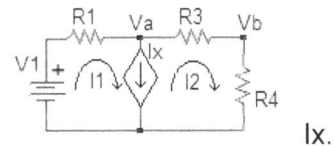

3. Use the value of Ix calculated in step 2 above and super mesh analysis to calculate the currents I1 and I2. Be sure to use your measured resistor values. Use your calculated currents, I1 and I2, to calculate the voltages, Va and Vb.

4. Calculate the percent difference between the calculated and measured voltages, Va and Vb.

Experiment 14: Node Voltage Analysis

Introduction

Node voltage analysis is based on Kirchhoff's current law: the algebraic sum of the currents at any node must be zero.   The example circuit below has two essential nodes with the voltages Va and Vb.

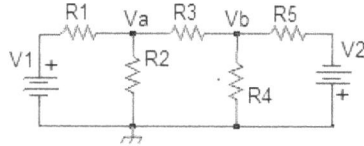

The application of Kirchhoff's voltage law results in the two independent simultaneous equations below.   Refer to your textbook or instructor for more details, if needed.

$$\frac{Va-V1}{R1}+\frac{Va}{R2}+\frac{Va-Vb}{R3}=0 \quad and \quad \frac{Vb-Va}{R3}+\frac{Vb}{R4}+\frac{Vb-(-V2)}{R5}=0$$

In the example above the values of the voltage sources and the resistances are all assumed to be known. The number of equations equals the number of unknowns.

In the example below a red light emitting diode (LED) is connected between two nodes. The voltage across an LED varies with the current through it. A typical red LED may have about 1.7V across it when its current is 5mA, and 2.0V across it when its current is increased to 20mA.

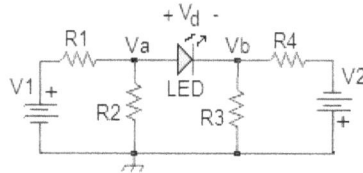

This circuit will be used as an example of the super node method. The nodes at Va and Vb will be taken as a super node. The super node equation sums the current away from each node as shown below.   The second equation relates the LED voltage to Va and Vb.

$$\frac{Va-V1}{R1}+\frac{Va}{R2}+\frac{Vb}{R3}+\frac{Vb-V2}{R4}=0 \quad and \quad Va=Vb+V_d$$

The diode voltage, V_d, is actually a logarithmic function of the diode current.   However, for the purpose of this experiment, we will measure V_d and use the measured value to calculate Va and Vb using the super node method.

Objectives

The primary objective of this lab experiment is to gain experience in the application of node voltage analysis.   Measurements will be made with the DMM and the results will be compared to theoretical calculations.   In part A node analysis will be applied to a resistor network.   In part B super node analysis will be applied to a circuit with a dependent source.

Procedure

> **Equipment and Parts**
>
> DMM, Power Supply, and Breadboard
> Resistors: 470, 750, 1.2k, two 1.5k, all ¼ watt, 5%.
> LED: Red (other colors may be used)

Part A: Node Voltage Analysis – Resistor Network

1. Measure the values of the resistors.

    R_{470} _____ R_{750} _____ $R_{1.2k}$ _____

    $R_{1.5k}$ _____ $R_{1.5k}$ _____

2. Connect the circuit on the right.   Layout the circuit in the same form as the diagram.   Do not use wires between the resistors.

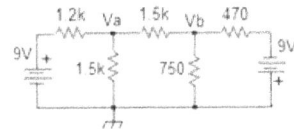

3. Set the plus and minus power supply voltages to exactly 9.0V.

4. Measure and record the voltages Va and Vb.

    Va _____   Vb _____

Part B: Super Node Analysis – Dependent Voltage Source

1. Connect the circuit on the right.   The LED may have one lead longer than the other.   The longer lead is usually the one that goes to the positive voltage (the diodes anode).

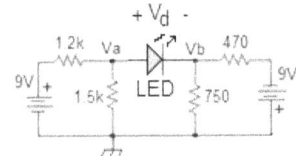

2. Measure and record the voltages Va and Vb.

    Va _____   Vb _____

3. Calculate and record the voltage V_d, as $V_d = Va - Vb$.

    V_d _____

Analysis, Part A

1. Use node voltage analysis to solve the circuit of part A for voltages Va and Vb. Use your measured part values to do the calculation.

2. Calculate the percent difference between the measured and calculated values of Va and Vb.

47

Analysis, Part B

1. Use the super node method to solve for the voltages Va and Vb using your measured part values.

    First write the equations summing currents away from the Va and Vb voltage nodes, ignoring the LED current.

    Next write an equation relating the voltages Va and Vb to the measured value of the LED voltage (refer to the introduction to this experiment).

    Solve the equations simultaneously for Va and Vb.

2. Calculate the percent difference between the calculated and measured voltages, Va and Vb.

3. Simulate your circuit using your part values. Use a voltage source (VDC) in place of the LED and set its value to the voltage that you measured across your LED.

History Note:

Edward Lawry Norton (July 29, 1898 - January 28, 1983)
Radio operator in the U.S. Navy between 1917 and 1919. Attended the University of Maine and M.I.T.   Worked for the Western Electric Company (predecessor to Bell Telephone Laboratories).   Received masters' degree in electrical engineering from Columbia University in 1925.   Worked for Bell Labs until retiring in 1963.

From Wikipedia, the free encyclopedia

Experiment 15: Network Theorems

Introduction

Thevenin's theorem states that a circuit between two nodes, a and b, can be replaced by a single voltage source (Thevenin voltage, V_{TH}) in series with a single resistance (Thevenin resistance, R_{TH}). The Thevenin voltage is equal to the "open circuit" voltage between nodes a and b. The Thevenin resistance is equal to the resistance between nodes a and b.

Norton's theorem states that a circuit between two nodes, a and b, can be replaced by a single current source (Norton current, I_N) in parallel with a single resistance (Norton resistance, R_N. The Norton current is equal to the "short circuit" current between nodes a and b. The Norton resistance is equal to the resistance between nodes a and b.

The application of Thevenin's or Norton's theorem may also require the application of other analysis methods. Determining an "open circuit" voltage or a "short circuit" current may require the use of node voltage or mesh current analysis.

Having a Thevenin or Norton equivalent circuit as a model of a more complex circuit makes it much easier to determine how the circuit will perform when loads or sources are connected to it.

It is possible to find the Thevenin or Norton equivalent of a circuit by measurement, without knowing, necessarily, the components and connections of the original circuit. If we can measure the original circuit's open circuit voltage and short circuit current, we can then determine the circuit's Thevenin or Norton resistance:

$$R_{Th} = R_N = \frac{V_{Th}}{I_N}.$$

Objectives

A bi-polar network is used to demonstrate the application of Thevenin's and Norton's Theorems. In part A, the Thevenin and Norton equivalent circuit will be determined by measurement. In part B, the Thevenin equivalent circuit will be connected and measurement results will be compared to the original circuit.

Procedure

Equipment and Parts
DMM, Power Supply, and Breadboard. Resistors: 1K, five 10K, all ¼ watt, 5%, 10KΩ trim-pot.

Part A: Bi-polar Network Circuit

1.   Measure the values of your resistors:

R1_____   R2_____   R3_____   R4_____

R5_____   1K_____

2. Connect the circuit shown on the right on a breadboard. Keep track of where your 10K resistors are. Layout the circuit on the breadboard in the same way it is in the schematic diagram.

Carefully note the power supply connections, especially that V1 is positive with respect to ground, and that V2 is negative with respect to ground.

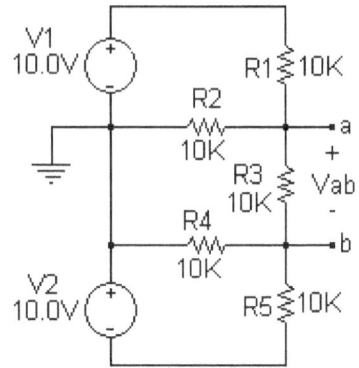

3. Measure and record the "open circuit" voltage Vab. Measure and record Va and Vb. (Vab = V_{TH})

Vab_____   Va_____   Vb_____

5. Measure and record the "short circuit" current between nodes a and b. Set the DMM to measure current on the 20mA range. If your meter shows less than 2 decimal digits after the decimal point, use a lower current range. Connect the positive meter lead to node a. Connect the negative lead to node b. Note that Iab = I_N, the Norton current.

Iab_____ I_N = _____

6. Calculate and record the Thevenin (and Norton) resistance: $R_{TH} = R_N = V_{TH} / I_N$.

R_{TH}_____

7. Connect a 1K "load" resistor between nodes a and b. Measure and record the "loaded" value of Vab.

Vab_{Loaded} _____

Part B: Thevenin and Norton Equivalent of Bi-polar Network Circuit

1. Connect the Thevenin equivalent circuit of the bi-polar network using the values of V_{TH} and R_{TH} from part A of this lab exercise. Use the 10K pot for R_{TH} by setting its value to that obtained in part A, step 4.

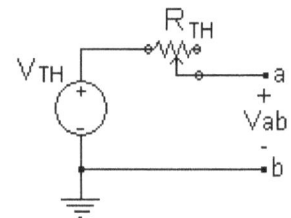

2. Set the power supply voltage to the value of V_{TH} you measured in part A of this lab exercise.

3. Connect the 1K load resistor that you used in part A of this lab exercise to your Thevenin equivalent circuit between nodes a and b.

4. Measure and record the value of Vab_{Loaded}. Vab_{Loaded}_____

5. Remove the 1k load resistor. Set the DMM to measure current on the 20mA range (same as in part A). Measure and record the "short circuit" current between nodes a and b by connecting the DMM between these nodes (positive lead to node a).

Iab_____   Note that Iab = I_N, the Norton current.

Analysis, Part A

1.   Write and solve the node voltage equations for Va and Vb for the bi-polar network. Be sure to use the measured values of your resistors. Calculate Vab.

2.   Use a spreadsheet to compare your calculated Va, Vb, and V_{TH} to your measured values. Calculate the percent error in the measurements compared to the calculations.

3.   Write and solve the mesh equations for the short circuit current, Iab = I_N, between nodes a and b. Use your spreadsheet to compare the result to your measurement by calculating the percent error. How much of the error would you attribute to meter loading effect?

Analysis, Part B

1.   Use your spreadsheet to compare the measurements of Vab, Vab_{Loaded}, and Iab in part B to the measurements in part A.

2.   Use your spreadsheet to compare the measurements in part B to the calculations (theoretical results) in part A.

TI-89 Example: Node Voltage Equations

Use Va = x and Vb = y for convenience. Also, the equations below use 10,000 Ω as the resistance of all of the resistors. Use your measured values in your equations.

Solve((x-10)/10000+x/10000+(x-y)/10000 = 0 and (y+10)/10000+y/10000+(y-x)/10000 = 0,{x,y})

x = 2.5, y = -2.5   Therefore Vab = V_{TH} = x – y = 2.5 + 2.5 = 5.0V.

TI-89 Example: Mesh Current Equations

Use I1 = x, I2 = y, and I3 = z for convenience. Also, the equations below use 10,000 Ω as the resistance of all of the resistors. Use your measured values in your equations.

Loop 1:   -10 + 10000 I1 + 10000 (I1 – I2) = 0
Loop 2:   10000 (I2 – I1) + 10000 (I2 – I3) = 0
Loop 3:   -10 + 10000 (I3- I2) + 10000 I3 = 0

solve(-10+10000*x+10000*(x-y)=0 and 10000*(y-x)+10000*(y-z)=0 and -10+10000*(z-y)+10000*z=0,{x,y,z})

x = I1 = 1mA, y= I2 = 1mA, z = I3 = 1mA.
Therefore I2 = I_N = 1mA.

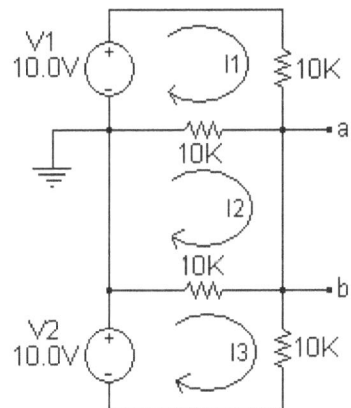

Experiment 16: Superposition and the Voltage Translator

Introduction

The superposition Theorem is particularly useful in analyzing circuits with multiple sources. This lab exercise involves a "voltage translator" circuit that has two power supply voltage sources, and a variable "signal" voltage source.

A signal source voltage varies from 0 to 6V. The output voltage of this translator varies from -0.9V to +0.9V. An input of 3V results in an output voltage of 0V.

The circuit in this exercise uses two fixed power supply voltages. In addition, it has a variable input voltage and a variable output voltage. The circuit translates the 0 to 6V "uni-polar" input voltage to a -0.9V to +0.9V "bi-polar" output voltage.

Objectives

Application of the superposition theorem involves analyzing the circuit one source at a time and summing (superimposing) the results. The theorem will be applied to the voltage translator circuit and its theoretical operation will be determined.

Procedure

Equipment and Parts
DMM, Power Supply, and Breadboard, Additional voltage source: 6V variable power supply. Resistors: R1 = 10K, R2 = 10K, R4 = 4.7K, all ¼ watt, 5%. R3 = 5K trim-pot.

1.   Measure and record the values of your resistors:

R1_____   R2_____   R4_____

2.   Connect the circuit shown below. Set Vin to 3.0V and adjust R3 for an output voltage of 0.0V. Measure and record V_{out} for each value of V_{in} given in the table below.

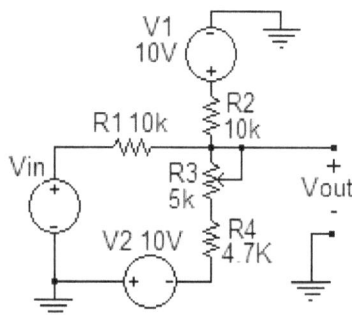

V_{in}	V_{out}
0.00	
1.00	
2.00	
3.00	
4.00	
5.00	
6.00	

3.   The following procedure will check out the superposition theorem by measurement. Note: If a supply voltage does not go down to 0.00V, disconnect the power supply and replace it by a short circuit (do not short circuit the power supply).

a)   Set V1 to 10.0V. Set V_{in} and V2 to 0.00V. Measure V_{out} and record the value in the table below.

b)   Set V2 to -10.0V. Set V_{in} and V1 to 0.00V. Measure V_{out} and record the value in the table below.

c)   Set V_{in} to 6.00V. Set V1 and V2 to 0.00V, Measure V_{out} and record the value in the table below.

Voltage Source	V_{out}
a) V1 only (+10V	$V_{out(1)} =$
b) V2 only (-10V	$V_{out(2)} =$
c) V_{in} only (+6V	$V_{out(in)} =$

4.   The superposition theorem states that: $V_{out} = V_{out(in)} + V_{out(1)} + V_{out(2)}$.
Check that the result for V_{out} from the table in step 3 above is close to the value of V_{out} in the table in step 2 for an input of 6.00V.

Analysis

1.   Write a node voltage equation for V_{out} given that V_{in} is a variable, V1 is +10V, and V2 is –10V. Write an equation for Vout using the measured values of your resistors. It should have the form: $V_{out} = mV_{in} - b.$ (find m and b).

2.   Use a spreadsheet to plot V_{out} from your data in the table of step 2 of the procedure. Compare the plot to the equation for V_{out} above (slope and intercept).

3.   Simulate your circuit using the measured values of your resistors. Compare the simulation results to the results in the table of step 2 of the procedure

History Note:

Michael Faraday, FRS (22 September 1791 – 25 August 1867) was an English chemist and physicist who contributed to the fields of electromagnetism and electrochemistry. Faraday studied the magnetic field around a conductor carrying a DC electric current. While conducting these studies, Faraday established the basis for the electromagnetic field concept in physics.

From Wikipedia, the free encyclopedia

53

Experiment 17: Function Generator and Oscilloscope

Introduction

A *function generator* is an instrument that produces a variety of waveforms whose amplitudes and frequencies can be varied. Typically, a function generator produces sine, square, and triangular waveforms. Frequencies may range from less than 1Hz to over 1MHz. The amplitude can usually be varied from close to zero to about 20V peak-to-peak

An *oscilloscope* is an instrument used mainly to observe waveforms in the time domain. In the "time domain" mode, it produces a graph of voltage versus time. An oscilloscope may also be operated in the "x-y" mode, where the voltage on the vertical axis can be plotted against the voltage on the horizontal axis.

Most oscilloscopes have two inputs which can be displayed simultaneously. The vertical axis is calibrated in volts per division and the horizontal axis is calibrated in seconds per division (in the x-y mode, the horizontal axis is also calibrated in volts per division).

Voltages or signals are amplified and scaled by "vertical amplifiers". A "time base" circuit scales the horizontal axis. A "trigger" circuit is used to obtain a stable display.

The main purpose of this lab is to learn how to use the function generator and oscilloscope. A *frequency counter* may be used to measure the frequency of the function generator. However, some function generators will display their exact frequency, so a frequency counter may not be needed.

The Oscilloscope

Modern oscilloscopes digitize the signals input to the vertical amplifiers, store the digitized signals in memory, process the information stored in memory, and display the results on an LCD screen. Digitizing is done by an "analog to digital converter". The vertical signal is sampled and converted to a sequence of binary words. For example, the Tektronix TDS1002 oscilloscope converts the input signal to a series of 8 bit "bytes" and stores 2500 bytes in memory as data. This data memory is processed and displayed on its LCD screen.

Most oscilloscopes have a variety of "modes" of operation. In the time mode (YT), the horizontal time scale is set by a *TIME/DIV* control. The vertical scale is set by the *VOLTS/DIV* controls. In the XY mode the horizontal scale is set by a *VOLTS/DIV* control. In the MATH mode signals from the vertical channels may be added, subtracted, or converted to another format.

A "trigger" circuit is used to obtain a stable display. If the waveform being displayed is periodic, each successive sweep can be made to coincide on the left side of the screen by using the trigger (*TRIG*) controls. The beginning of each sweep is delayed until the waveform reaches the exact same position in the cycle so that each sweep produces an identical picture.

If the waveform is not triggered properly, the display is unsynchronized and unstable. You must select the proper trigger source to match the waveform being displayed and you can adjust the trigger *LEVEL* and *SLOPE* to begin the waveform where desired.

Simplified Block Diagram of a Digital Oscilloscope

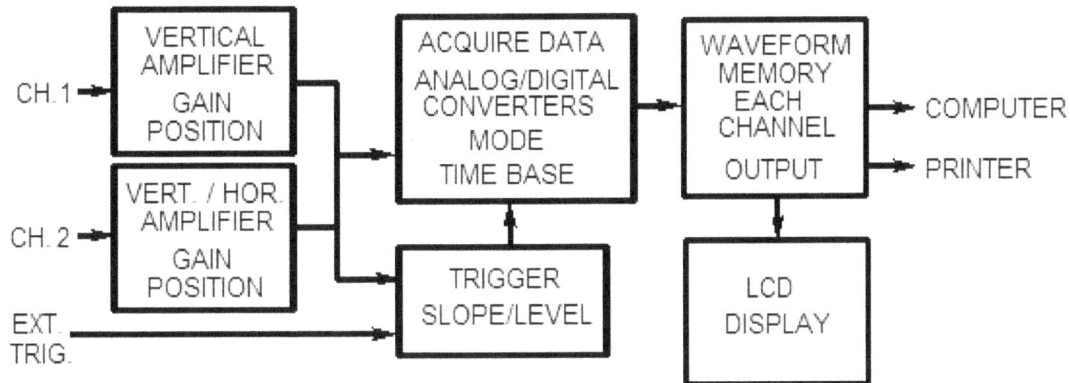

The block diagram above shows the functional blocks of a typical digital oscilloscope. One needs to understand the function of each block and the signal paths in the diagram to be able to use the oscilloscope effectively.

Each vertical channel has three basic controls: *VOLTS/DIV, VERTICAL POSITION,* and *INPUT COUPLING.* In XY mode, channel 2 becomes a horizontal amplifier.

A *VOLTS/DIV* control scales the display by setting the gain of the vertical / horizontal amplifier. The vertical axis of the oscilloscope usually has 8 major divisions. For example, the Tektronix TDS1002 vertical amplifier gain can be set from 2 millivolts per division to 5 volts per division.

A *VERTICAL POSITION* control positions the displayed signal vertically.

An *INPUT COUPLING* control sets the vertical input to AC, DC, or GND. In the AC mode, only AC signals are displayed. In the DC mode, both AC and DC signals are displayed. Selecting GND connects the input to ground so that no signal is displayed.

The *TIME BASE* sets the calibration of the horizontal axis. The horizontal axis of the oscilloscope usually has 10 major divisions. For example, the Tektronix TDS1002 time base can be set from 5 nanoseconds per division to 50 seconds per division. The display position can also be adjusted horizontally.

The *TRIGGER* circuit is used to get a stable display. The trigger source may be a signal from channel 1, channel 2, or an external source. The trigger point on the waveform is adjustable with the *LEVEL* and *SLOPE* controls. The trigger source may also be filtered.

Digital oscilloscopes usually have a means of connecting to a computer through a USB, GPIB, Ethernet, or serial port. Data from the oscilloscope can be downloaded to the computer directly into spreadsheet programs.

Also the oscilloscope screen can be captured and copied to the computers clipboard, or sent directly to a printer. The specific data acquisition capability depends on the oscilloscope and available software.

Note: The instructor may supplement this lab exercise with specific information on the types and models of instruments at your lab station.

Function Generator

The function generator typically generates sine waves, square waves, and triangle waves. The "frequency" and "amplitude or level" of these waves can be adjusted, and a "DC offset" can be added. The generator may have more than one output. The main output should go to your circuit. Another output may be provided which can be used to connect it to an oscilloscope trigger circuit or to a frequency counter.

Quality laboratory function generators use digital waveform synthesis to generate the output waveforms. Older analog function generators may not have an accurate frequency display so a frequency counter would be needed to set the frequency accurately.

Depending on the particular function generator, the frequency may be set using a *RANGE* switch and a variable *FREQUENCY* control, or by entering the frequency with a keypad. The type of waveform is set using the *WAVEFORM* switch (or button, or keypad).

A provision may be available to offset the waveform so that it is not centered on zero volts. An *OFFSET* control varies the amount of offset. The offset is adjusted by observing the waveform on the oscilloscope. The oscilloscope must be set for *DC* input so that both the DC and AC components of the waveform are observed. The *AMPLITUDE* or *LEVEL* control varies the output level. Some function generators may be set up entirely by means of a keypad, or remotely by a computer.

Procedure

Equipment and Parts
Oscilloscope, Function Generator, and 100Ω, 5% resistor

Part A: Basics

1. Connect the instruments as shown below. Note that the connections are made with "coaxial cables" with "clips" at their ends. The oscilloscope clip may be part of a "probe". Connect the clips as shown below. The clip connected to the ground lead is usually color coded black.

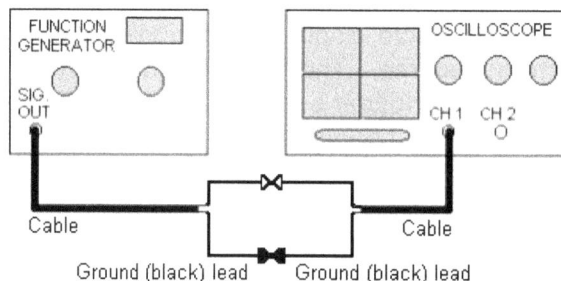

2. Turn on the oscilloscope, set the *TRIGGER MODE* to "Auto", and establish a single trace (a horizontal line). Set input for channel 1 only, and set the *INPUT COUPLING* to *GND* (the input coupling switch may be under or near the *VOLTS/DIV* switch).

   You should see a straight horizontal line (the trace). Adjust the *VERTICAL POSITION* control to center the trace on the screen.

3. Set the channel 1 *INPUT COUPLING* switch to *DC,* and the *VOLTS/DIV* switch to 1V/DIV. Set the *TIME/DIVISION* switch to 10μS/DIV.

4. Set the function generator to produce a 4 volt peak-to-peak square wave without a dc offset. Set the frequency to exactly 10kHz.

5. Set the oscilloscope's trigge*r SLOPE* control to "negative". Adjust the trigger *LEVEL* control for a stable trace, like the one shown below. Notice that the trigger event occurs at the center of the screen. The display may be moved horizontally with the *HORIZONTAL POSITION* control.

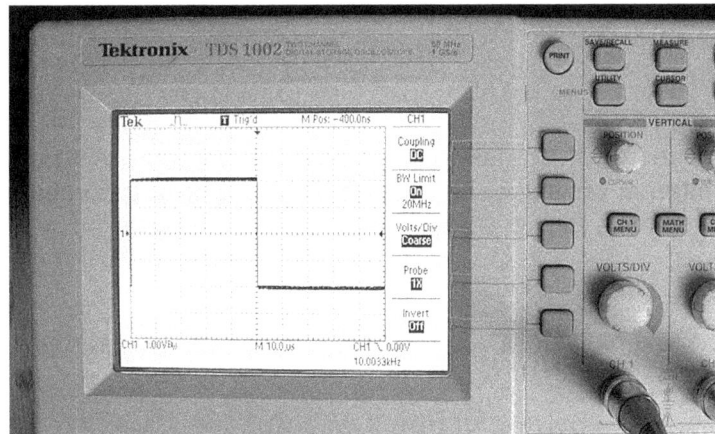

Note the easy to use menu system. The channel 1 menu appeared on the right side of the screen when the *CH1 MENU* button was pressed. The buttons to the right of the screen allow easy selection of channel 1 settings. Channel 2 menu, Trigger menu, and horizontal menus work the same way.

6. The period of a 10kHz waveform is 100 microseconds (T = 1/f). The oscilloscope's time base is set to 10μS/DIV. Since there are 10 horizontal divisions, the total time displayed is 100μS. Therefore there is one cycle displayed on the screen (100μS per cycle).

    The amplitude display of the function generator may not agree with the amplitude displayed on the oscilloscope. Always use the displayed oscilloscope amplitude as the correct amplitude.

7. Change the time per division on the oscilloscope to 50μS/DIV. How many cycles are displayed?

    Number of cycles = _____

Part B: Generator Source Resistance

1. The function generator can be modeled as a Thevenin source, as shown below.

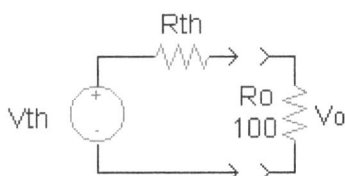

Rth = internal resistance of function generator

Vth = function generator output with no load

Vo = function generator output with a load

57

2. In part A, you set the function generator to produce 4V peak-to-peak without a "load" resistor connected. Therefore Vth is 4V peak-to-peak since there is no current through Rth and therefore no voltage drop across it. Connect the 100Ω resistor, Ro, to the function generator without changing any of the settings.

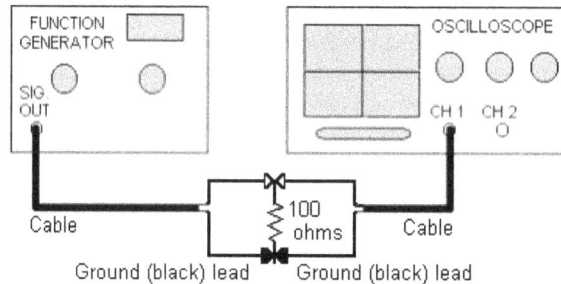

Measure and record the peak-to-peak voltage, Vo, across the 100Ω resistor. If the generator's internal resistance is about 50Ω (most are), the voltage should decrease to about two thirds of its former value (2.67V).

Vo = _____

3. Remove the 100Ω resistor. Calculate the generator's internal resistance, Rth:

$$Vo = \frac{Ro}{Ro + Rth} Vth \ . \ \text{Solving for Rth:} \quad Rth = Ro\left(\frac{Vth}{Vo} - 1\right)$$

Rth = _____

Part C: Settings Exercise

1. Change the function generator waveform to a sine wave. Adjust trigger slope and trigger level controls to get the display shown on the right.
(f = 10kHz, v = 4V peak-to-peak)

The display on the right was obtained with the trigger slope negative, and trigger level set to zero.

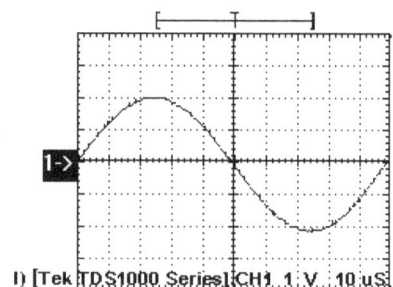

2. Note and explain what happens when:

a)	Trigger slope is changed to positive (rising). Where is the trigger event on the screen? Why does the waveform's appearance change?

b)   Trigger level is changed to +1V. Where is the trigger event on the screen? Why does the waveform's appearance change?

c)   Time per division is changed to 100µS/Div and trigger level to 0V. Why does the waveform's appearance change?

d)   Volts per division is changed to 2V/Div. Why does the waveform's appearance change?

3.   Reset the oscilloscope and function generator to the settings of part C, step 1. Be sure that the oscilloscope input is set to "DC" coupling.

4a.   Add a +1.00V offset to the function generator's waveform. Most function generators will have a button or knob labeled "offset voltage" (you may need to read the instrument's instruction manual),

4b.   Adding a +1V offset to the waveform should cause it to move vertically on the oscilloscope by 1V. When the oscilloscope's input coupling is changed to AC, a capacitor is connected in series with the input. The capacitor blocks the DC component of the signal. Note what happens when the oscilloscope's input coupling is set to "AC".

Explain your observations below:

a.   +1.00 volt offset is added. Why does the waveform's vertical position change?

b.   Input coupling is changed to AC. Why does the waveform's vertical position change?

Lab Check

Your lab instructor will indicate what is required to get credit for this lab exercise. Possibilities include:

1.   Each lab section may be checked and initialed as the exercise is performed.

2.   Entire lab may be checked when the exercise is completed.

3.   A report may be required describing the outcomes of each section.

The lab instructor should be consulted if there is any uncertainty about any of the measurements or instrument settings while doing this lab exercise.

Experiment 18: RC Transient Response

Forced Step Response

Consider the circuit on the right. Before t = 0, the capacitor is discharged and its voltage, v_C, is zero. When the switch is closed at t = 0, the capacitor charges through the resistor and its voltage increases exponentially with time.

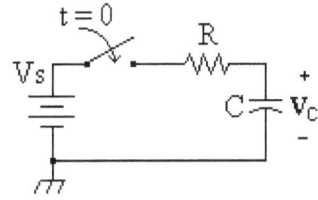

The response of this circuit can be divided into two parts, the "transient response" and the "steady state response".

The transient part occurs while the capacitor is charging and its voltage is changing. The steady state part occurs when the capacitor is fully charged and its voltage stops changing.

This transient response is called a "forced step" response, or just "step response", because the voltage applied to the circuit changes value instantly. The capacitor voltage, v_C is given below as a function of time.

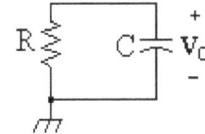

$$v_C(t) = V_S(1 - e^{\frac{-t}{RC}}) + v_C(0)$$

The voltage $v_C(0)$ is the capacitor voltage at t = 0. An important parameter of the RC circuit is the "time constant": $\tau = RC$.

Natural Response

Natural response refers to the response of the RC circuit when the circuit is discharging its stored energy. The amount of energy stored in the capacitor can be calculated using the equation below. In the SI system, energy, W, has units of joules, voltage, V, has units of volts, and C has units of farads.

$$\text{Capacitor energy: } W = \frac{1}{2}CV^2$$

The capacitor voltage, v_C is given below as a function of time.

$$v_C(t) = V_i e^{\frac{-t}{RC}} \qquad \text{where the initial capacitor voltage, } v_C(0) = V_i$$

Objectives

The step and natural response of an RC circuit will be measured and compared to calculations. A "square wave" waveform will be used. The generator voltage will alternate between 0 V for 50μS and 8V for 50μS. The voltage stays constant long enough for the circuit to reach "steady state".

When the voltage changes from 0V to 8V, the circuit's step response will be observed. When it changes from 8V to 0V, the circuit's natural response will be observed.

Procedure

> **Equipment and Parts**
>
> Oscilloscope, Function Generator, and Breadboard
> Resistor: 680Ω, ¼ watt, 5%.   Capacitor: 10nF, 5%

1. Connect the instruments as shown.

2. Set the function generator to produce a square wave. Set the *FREQUENCY* to 10kHz.

    Adjust the *AMPLITUDE* and *OFFSET* controls to obtain a 0V to 8V square wave.

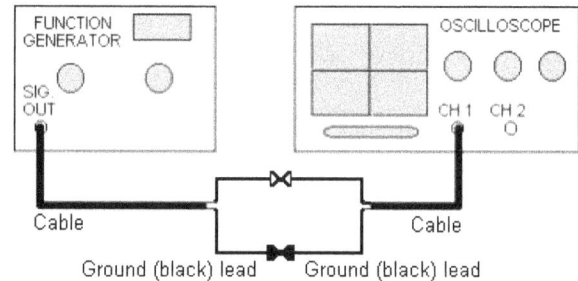

3. Turn on the oscilloscope. Set the *TRIGGER MODE* to "*Auto*". Set the input for channel 1 only, and set the *INPUT COUPLING* to *GND*. The input coupling switch is usually right under, or near, the *VOLTS/DIV* switch, or use the *CH1 MENU* button.

4. Adjust the *VERTICAL POSITION* control so that the trace starts at the bottom of the screen. Set the *INPUT COUPLING* switch to *DC,* and the *VOLTS/DIV switch* to 1 VOLT/DIV. Set the *TIME/DIV* switch to 10µS/DIV.

5. Set the oscilloscope *TRIGGER SLOPE* to negative (falling edge) and *TRIGGER LEVEL* controls to obtain a stable trace. Waveform should begin at the top of the screen (+8V) and after 50µS (5 horizontal divisions) it should go to the bottom of the screen (0V).

6. Connect the RC circuit to the function generator as shown below. Be sure to leave channel 1 of the oscilloscope connected to the function generator.

    The function generator is shown below as a Thevenin equivalent circuit. Vth is the "no load" generator voltage and Rth is the generator's internal resistance. A common value for Rth is 50Ω. Connect channel 2 of the oscilloscope so that it is across the capacitor to measure Vc, as shown below (display shown is for 5V peak-to-peak, yours will be 8V peak-to-peak).

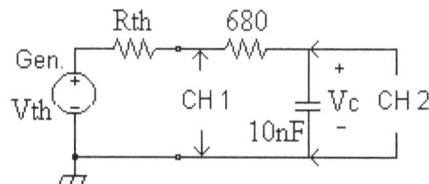

7. Set the oscilloscope to display channel 1 and channel 2 as shown on the right. The left 5 horizontal divisions display the "forced" response of the circuit the right 5 horizontal divisions display the "natural" response.

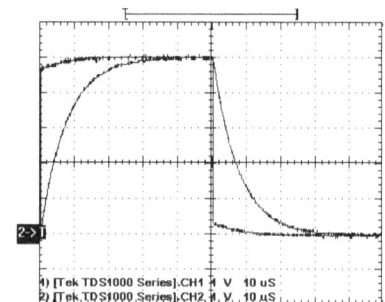

    Notice that the channel 1 waveform has a small "dip" in it due to the function generator's internal resistance.

62

When the function generator's output goes to 0V, its internal resistance is still in the circuit. It must be accounted for when calculating the circuit's time constant: $\tau = RC = (Rth + 680)(10nF)$.

8. Change the oscilloscope's *TIME/DIV* to 5µS/DIV. You should see only the forced response. You may need to adjust the *HORIZONTAL POSITION* control to get the display shown below on the right.

    Record the time of occurrence of the predicted voltage for 1, 2, 3, and 4 time constants after t = 0. Use the spreadsheet format given at the bottom of this page. Also see the example below for calculating the forced response.

    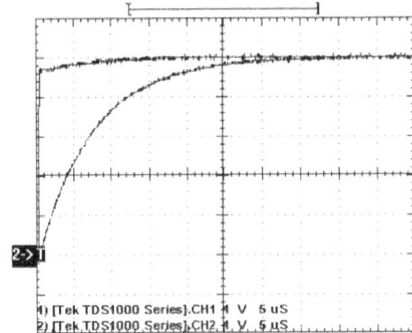

    $$v_C(1\tau) = 8(1 - e^{-1}) = 8(1 - .368) = 5.06 \; volts$$

    $$v_C(2\tau) = 8(1 - e^{-2}) = 8(1 - .135) = 6.92 \; volts$$

    $$v_C(3\tau) = 8(1 - e^{-3}) = 8(1 - .098) = 7.60 \; volts$$

    $$v_C(4\tau) = 8(1 - e^{-4}) = 8(1 - .018) = 7.85 \; volts$$

    The horizontal position control has to be adjusted so that the trigger occurs on the left side of the screen to get the display shown above on the right (display shown is for 5V peak-to-peak, yours will be 8V peak-to-peak).

9. Repeat step 8 above for the circuit's natural response. Keep the oscilloscope's *TIME/DIV* at 5 µSec/DIV, but change the trigger to observe the natural response.

	A	B	C	D	E	F	G	H	I	J	K	L	M
1	RC Forced:							RC Natural:					
2	Tau	Voltage (V)	Meas. Time (us)	Calc. Time (s)	% Error	Current (A)		Tau	Voltage (V)	Meas. Time (us)	Calc. Time (s)	% Error	Current (A)
3	1	5.06						1					
4	2	6.92						2					
5	3	7.60						3					
6	4	7.85						4					

Analysis

1. Calculate the theoretical time constant values and enter the values into the spreadsheet columns D an K. Calculate the percent error values in columns E and L. You can enter the appropriate equations into the cells in row 4 and use "fill down".

2. Calculate the circuit's current with the correct polarity for the circuit's step and natural response and put the result into columns F and M. Explain when (or at which multiple of tau) and why the maximum (absolute value) currents occurred.

3. Simulate the circuit and plot the capacitor voltage and current as a function of time. Refer to the simulation example on the next page.

4.   Compare the measured step and natural response time constants to the simulation results. How do these compare to the calculation: $\tau = (Rth + R)\, C$?

        Lab Report: Include spreadsheet, simulation results, and answers to above questions.

LTspice Example: RC Transient Response

V1 is set by clicking the right mouse button on it and clicking "advanced" in the dialog box shown below.

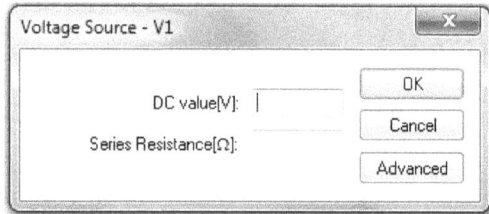

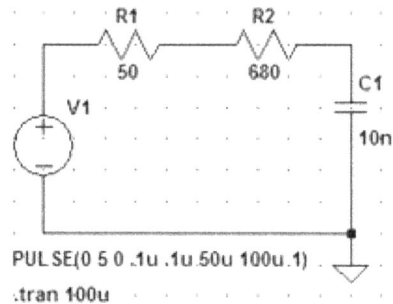

PULSE(0 5 0 .1u .1u 50u 100u .1)

.tran 100u

This will open the dialog box shown on the right. Select *PULSE*.

Enter the desired pulse characteristics. *LTspice* pulse source also has an "*Ncycles*" parameter.

Select transient analysis with a stop time of 100µS and maximum time step of 0.1µS.

Placing the mouse pointer over a part After running the simulation displays an arrow which shows the direction of positive current.

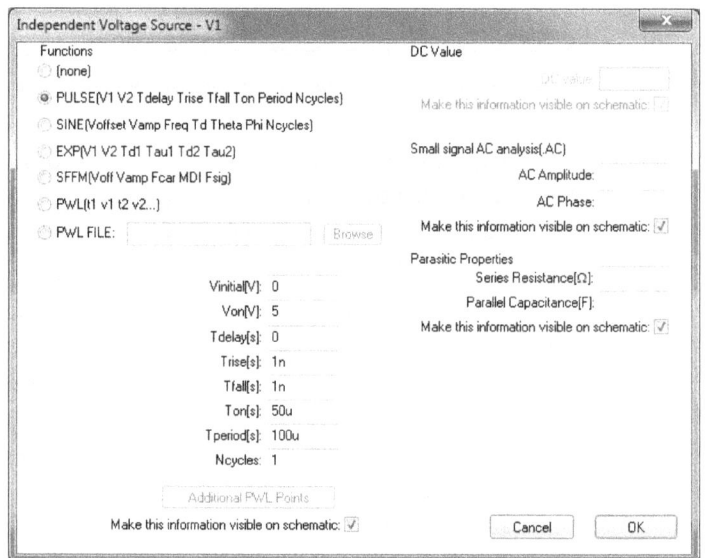

The graph below on the left shows the capacitor voltage and the voltage between R1 and R2. Click the probe on the nodes where you wish to see the voltage. The graph below on the right shows the capacitor current. Move the mouse over the capacitor and click to display the current waveform.

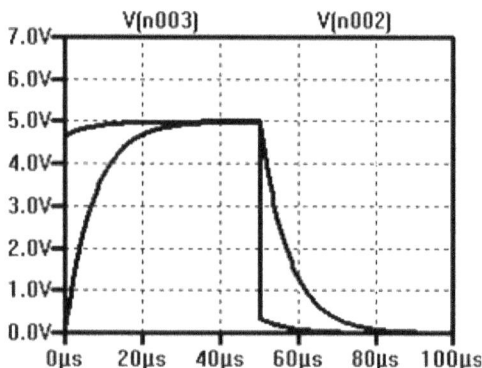

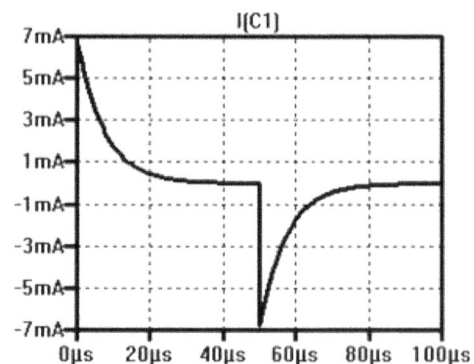

Experiment 19: RL Transient Response

Forced Response

The properties of "inductance" and "inductors" are complementary to the properties of "capacitance" and "capacitors". Inductors store energy in a magnetic field while capacitors store energy in an electric field. Inductors have an inertia to a change in current while capacitors have an inertia to a change in voltage.

An RL circuit's <u>current response</u> to a step voltage is of the same form as the RC circuit's <u>voltage response</u> to a step voltage. Also, the RL circuit's natural current response is of the same form as the RC circuits natural voltage response.

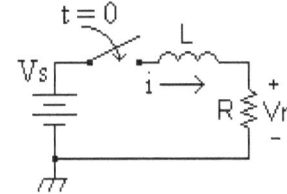

The equation for the current as a function of time for the circuit on the right is given by:

$$i(t) = I_f \left(1 - e^{\frac{-Rt}{L}} \right)$$

The current I_f is the circuit current at t = ∞ , or I_f (∞) (a final condition). Initial current is 0.

An important parameter of the RL circuit is the time constant: $\tau = \dfrac{L}{R}$.

Natural Response

The amount of energy stored in the inductor can be calculated using the equation below. In the SI system, energy, W, has units of Joules, current, I, has units of Amperes, and inductance, L, has units of Henries.

$$\text{Inductor energy}: \quad W = \frac{1}{2}LI^2$$

I_i is the inductor current at t = 0 in the circuit on the right, or i(0).

Inductor current, i(t), for t ≥ 0 is given by: $\quad i(t) = I_i e^{-\frac{Rt}{L}}$

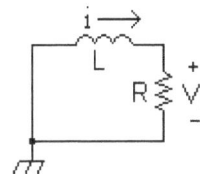

Objectives

The step and natural response of an RL circuit will be measured and compared to calculations. The generator voltage stays constant long enough for the circuit to reach "steady state" (about 5 time constants in this experiment).

When the voltage changes from 0 V to 8V, the circuit's forced response will be observed. When it changes from 8V to 0V, the circuit's natural response will be observed.

Procedure

Equipment and Parts
Oscilloscope, Function Generator, and Breadboard Resistors: 1000Ω, 5%. Inductor: 10mH, 5%, air core (J.W. Miller 70F102AI).

1.   Connect the instruments shown.
     Set the generator to 10kHz. Adjust the AMPLITUDE and OFFSET controls to obtain a 0 to 8V square wave.

2.   Turn on the oscilloscope. Set the TRIGGER MODE to AUTO.

3.   Set the input for channel 1 only, INPUT COUPLING to GND.

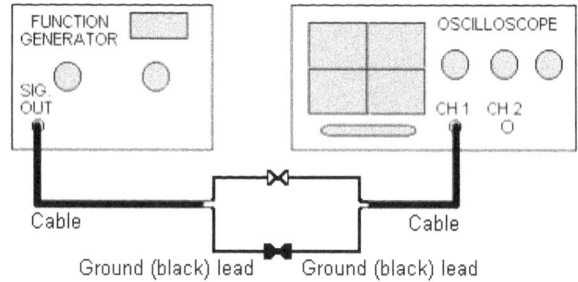

     You should see a straight line on the screen. Adjust the VERTICAL POSITION control to center the trace on the screen vertically.

4.   Re-adjust the oscilloscope's VERTICAL POSITION control so that the trace fits between the bottom and top of the screen.

     Set the INPUT COUPLING switch to DC, and the VOLTS/DIV switch to 1V/DIV. Set the TIME/DIV switch to 10µSec/DIV.

5.   Set the oscilloscope TRIGGER SLOPE to negative (falling edge) and TRIGGER LEVEL controls to obtain a stable trace.

     Waveform should begin at the top of the screen (+8V), and after 50µS (5 horizontal divisions) it should go to the bottom of the screen (0V).

6.   Connect the RL circuit to the function generator as shown below on the right. Be sure to leave channel 1 of the oscilloscope connected to the function generator.

     The function generator is shown below as a Thevenin equivalent circuit. Vth is the "no load" generator voltage and Rth is the generator's internal resistance.

     Connect channel 2 of the oscilloscope so that it is across the 1K resistor to measure Vr, as shown below.

     Rx is the DC resistance of the inductor, which must be measured with an ohmmeter.
     Measure and record the value of Rx.

     Rx: _____

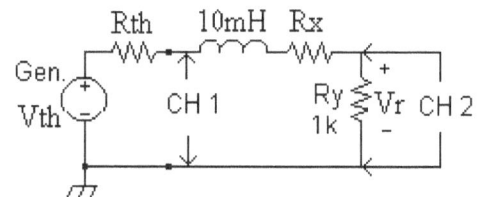

Your display should look like the one on the right, except the amplitude will be 8V peak-to-peak. Notice that the channel 1 waveform voltage is greater at the left side and decreases toward the center. Can you explain this?

Channel 2 displays the "forced" response of the circuit in the left 5 major scale divisions, and the "natural" response in the right 5 major divisions.

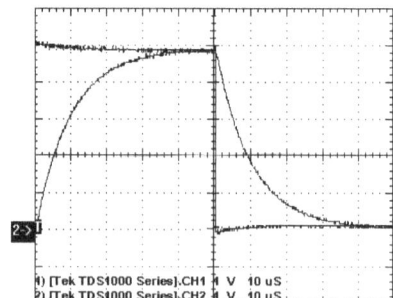

66

7. Change the *TIME/DIV* to 5µSec/DIV. Adjust the horizontal position and trigger to observe the forced response only.

. The current in the RL circuit can be calculated from the voltage drop, Vr, across the 1K resistor. First measure and record the steady state voltage, V_f, at t = 50µSec. Since the circuit current is proportional to the voltage, Vr, the value of Vr can be calculated as a function of the time constant.

V_f = _____

Calculate the expected voltage for each time interval below and record in the spreadsheet shown at the bottom of page in column B. The example below is for the forced response.

$$1\tau = V_f\left(1 - e^{-1}\right) = V_f(1 - .368) = \underline{\hspace{2cm}} \text{ volts}$$

$$2\tau = V_f\left(1 - e^{-2}\right) = V_f(1 - .135) = \underline{\hspace{2cm}} \text{ volts}$$

$$3\tau = V_f\left(1 - e^{-3}\right) = V_f(1 - .050) = \underline{\hspace{2cm}} \text{ volts}$$

$$4\tau = V_f\left(1 - e^{-4}\right) = V_f(1 - .018) = \underline{\hspace{2cm}} \text{ volts}$$

Record the time of occurrence of the predicted voltages for 1, 2, 3, and 4 time constants after t = 0. Record in spreadsheet column C

8. Repeat step 7 above for the circuit's natural response. Keep the oscilloscope's *TIME/DIV* at 5µSec/DIV, but change the trigger to observe the natural response.

	A	B	C	D	E	F	G	H	I	J	K	L	M
1	RC Forced:							RC Natural:					
2	Tau	Voltage (V)	Meas. Time (us)	Calc. Time (s)	% Error	Current (A)		Tau	Voltage (V)	Meas. Time (us)	Calc. Time (s)	% Error	Current (A)
3	1							1					
4	2							2					
5	3							3					
6	4							4					

Analysis

1. Calculate the theoretical time constant values and enter the values into the spreadsheet columns D an K. Calculate the percent error values in columns E and L. You can enter the appropriate equations into the cells in row 4 and use "fill down.

2. Calculate the circuit's current with the correct polarity for the circuit's step and natural response and put the result into columns F and M. Explain when (or at which multiple of tau) and why the maximum (absolute value) currents occurred.

3. Simulate the circuit and plot the inductor voltage and current as a function of time. Refer to the simulation example at the end of this experiment.

4.   Compare the measured step and natural response time constants to the simulation results. How do these compare to the calculation: $\tau = L/(Rth + Rx + Ry)$?

Lab Report: Include spreadsheet, simulation results, and answers to above questions.

LTspice Simulation Example

The simulation settings for this circuit are the same as for the RC circuit in the previous lab exercise. Note the effect of the function generator's source resistance on the square wave at node N1.

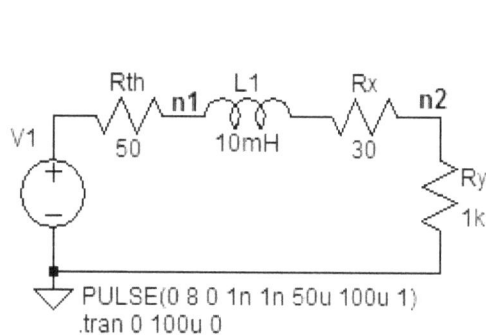

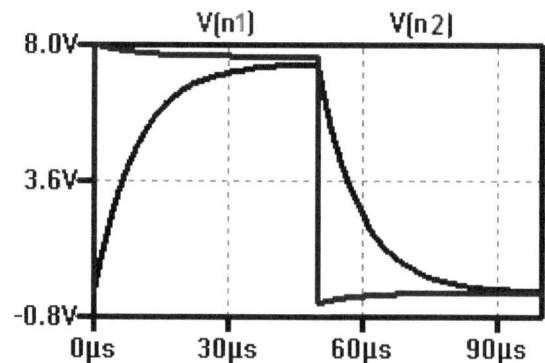

History Note:

Joseph Henry (17 December 1797 – 13 May 1878) an American scientist. Henry discovered the electromagnetic phenomenon of self-inductance. He also discovered mutual inductance independently of Michael Faraday, though Faraday was the first to publish his results. The SI unit of inductance, the henry, is named in his honor. Henry's work on the electromagnetic relay was the basis of the electrical telegraph, invented by Samuel Morse and Charles Wheatstone separately.

From Wikipedia, the free encyclopedia

Experiment 20: Capacitor Network Transient Response

Introduction

Series connected capacitors receive the same current when charged. Each capacitor will have the same charge Q, regardless of the value of its capacitance. However the voltage across each capacitor will depend on its capacitance.

Consider the circuit on the right. Capacitor voltage, V1 + V2, will be close to 10V if the switch is closed for more than 5 time constants.

Given that C1 = 10nF and C2 = 40nF, find V1 and V2:

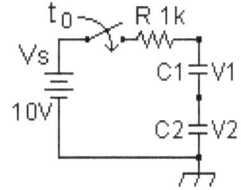

$$C_T = \frac{10 \cdot 40}{10 + 40} nF = 8nF \qquad Q_T = Q_1 = Q_2 = C_T \cdot Vs = (8nF) \cdot (10 \text{ Volts}) = 80nC$$

$$V_1 = \frac{Q_1}{C_1} = \frac{80nC}{10nF} = 8V \qquad V_2 = \frac{Q_2}{C_2} = \frac{80nC}{40nF} = 2V$$

The RC time constant of the circuit is determined by C_T and R which is equal to 80 microseconds in this case. This lab experiment will show that all of the capacitors in the capacitor network charge at the same rate which is determined by the net capacitance of the network, C_T, and the circuit resistance, R.

Procedure

> ### Equipment and Parts
>
> Function Generator, Oscilloscope, and Breadboard.
> 10nF, 22nF, 5% capacitors, 1k resistor, ¼ watt, 5%.

1. Connect the circuit on the right.

2. Set the generator to produce a 5kHz, 10V peak-to-peak square wave with a plus 5V offset.

3. Connect oscilloscope channel 1 to node n1. Trigger on channel 1.

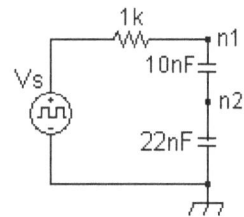

    Connect oscilloscope channel 2 to node n2. Set the oscilloscope time base to display one cycle of the response.

    Both channels should be set to DC input coupling.

4. Adjust the oscilloscope display so that you can accurately measure the response time constants, T_1 and T_2, and steady state peak-to-peak amplitudes of, V_1 and V_2, at nodes n1 and n2 (steady state amplitude occurs after 5 time constants when the capacitor is fully charged). Record below:

    T_1: _____ V_1: _____

    T_2: _____ V_2: _____

Analysis

1. Since the capacitors are in series, each capacitor has the same amount of charge. Also, the amount of charge on the series combination of the two capacitors is the same as the charge on each capacitor. Calculate the theoretical value of the steady state charge on the capacitors.

2. Use the results of step 4 to calculate the steady state charge on the 10nF capacitor (five time constants after Vs goes from 0 V to 10V) in coulombs. Use the voltage across the 10nF capacitor and Q = CV. Compare your results with those in analysis step 1

3. Use the results of step 4 to calculate the steady charge on the 22nF capacitor. Compare your results with those in analysis step 1.

4. Explain why the oscilloscope waveform on channel 2 goes negative (hint below).

5. Calculate the circuit's time constant, and compare the result to your measurement. How do the time constants for the 10nF and 22nF capacitors compare and why?

6. Simulate the circuit and compare the simulated results to your measurements. Explain any differences.

LTspice Simulation Example

Refer to the command lines below the schematic for the simulation settings (PULSE and .tran).

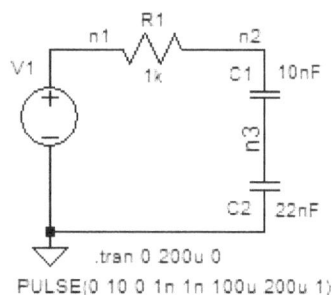

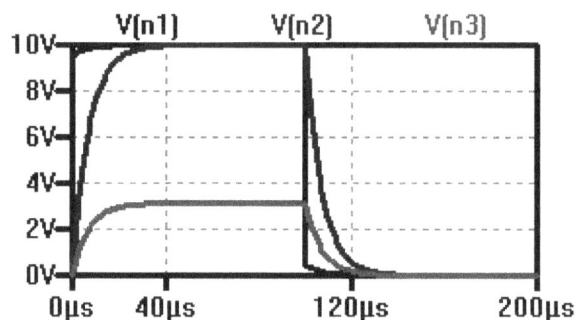

a) Extend simulation time to 50 cycles (.tran Stop Time: 10000u and PULSE Ncycles: 50). Observe the waveform at node n3.

b) Connect a 1MΩ resistor in parallel with C2 to simulate the loading effect of the scope's 1X probe. Observe the waveform at node n3. See analysis step 4.

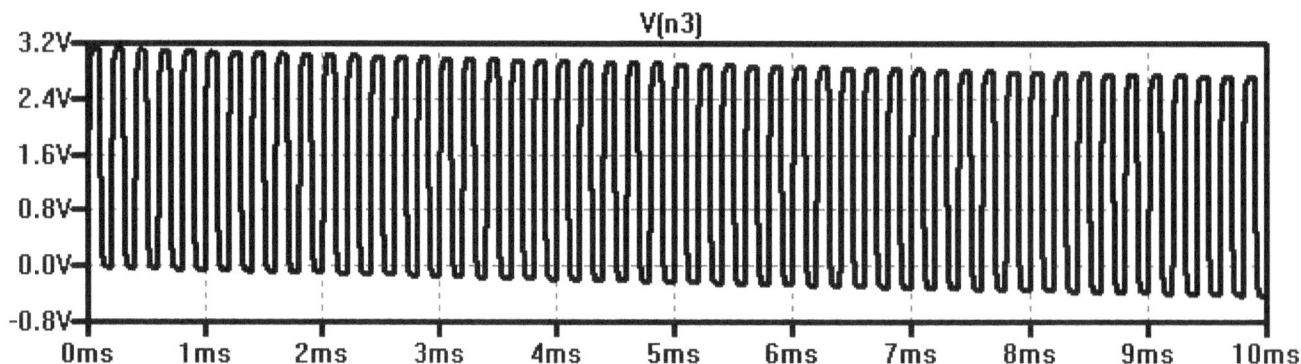

70

Experiment 21: Superposition of AC and DC Voltages

Introduction

Superposition of AC and DC voltages is very common in electronic circuits. The AC component may be a signal carrying information, an AC power line voltage, or it may be an undesired "noise" voltage. The DC component may be a voltage necessary for the operation of a transistor amplifier, or the output of a DC power supply.

This lab exercise investigates the effect of voltage dividers and "coupling" capacitors on AC and DC voltages. A coupling capacitor will pass an AC voltage, but block a DC voltage. The capacitor does provide opposition to current flow similar to a resistor. The capacitor's "resistance" is frequency dependent and is called "capacitive reactance."
The value of its capacitive reactance is given by:

$$X_C = \frac{1}{2\pi f C}$$ C is capacitance in Farads and f is frequency in Hertz.

Procedure

Equipment and Parts
Oscilloscope, Function Generator and Breadboard. Resistors: R1=1K, R2=1K, and R3=1.2K, all ¼ Watt, 5%. Capacitor: 100nF, 5%.

Part A: DC Measurements

1.   Measure the resistors: R1: _____   R2: _____   R3: _____

2.   Connect the circuit on the right. Connect oscilloscope channel 1 to node N1 and channel 2 to node N2.

3.   Set the coupling on both channels to "GND". Set the trigger to channel 1 and to "auto." Set both traces to the center of the screen with the vertical position controls.

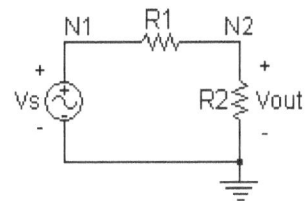

Set channel 1 and channel 2 to 1V/DIV.
Change the coupling on both channels to DC.
Set the time base to 2.5µS/DIV (or 2µS/DIV).

Set the generator to produce a 3.0V peak-to-peak, 100kHz sine wave with a +1.5V offset. This 1.5V offset will act like a 1.5V DC source in series with the 3V peak-to-peak waveform.

Important: Use the oscilloscope to set the amplitude and offset of the sine wave. Your oscilloscope display should be similar to the one shown on the right.

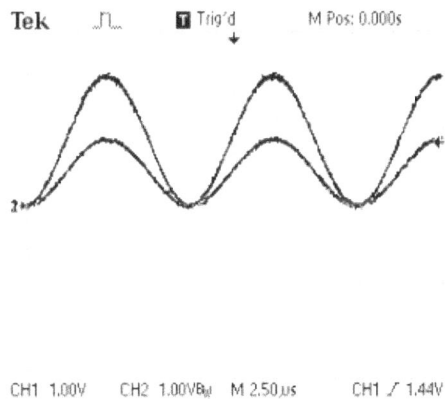

CH1 1.00V   CH2 1.00VBw   M 2.50µs   CH1 ⌁ 1.44V

The function generator with a DC offset voltage can be represented by a DC voltage source in series with an AC voltage source as shown on the right.

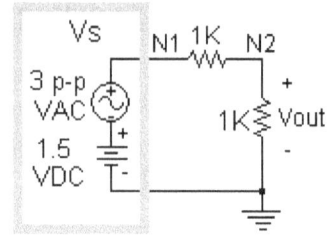

4. Make sure the DMM is in DC mode and measure the DC voltages at nodes N1 and N2. Record below:

    VDC_{N1}: _____ VDC_{N2}: _____

5. Change the coupling on both oscilloscope channels to AC. This removes the DC component from the oscilloscope so that it only measures the AC component. Both waveforms should be centered on 0.0 volts.

6. Change both vertical channels to 0.5V/DIV. Measure the peak-to-peak AC voltages at nodes N1 and N2 with the oscilloscope. Record below:

    VAC_{N1}: _____ VAC_{N2}: _____

Part B: AC Superposition

1. Connect the circuit on the right. Connect channel 1 to node N1 and channel 2 to node N3.

2. Set the coupling on both channels to "GND". Set the trigger to channel 1 and to "auto." Center both traces.

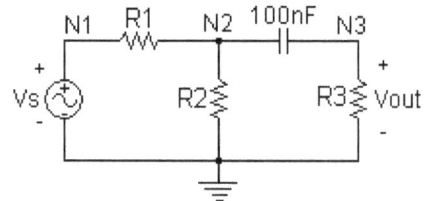

    Set channel 1 and channel 2 to 1V/DIV.
    Set the coupling on both channels to DC.
    Set the timing to 2.5µS/DIV (or 2µS/DIV).

    Set the generator to produce a 3.0V peak-to-peak, 100kHz sine wave with a +1.5V offset.

    Important: Use the oscilloscope to set the amplitude and offset of the sine wave. Your display should be similar to the one on the right.

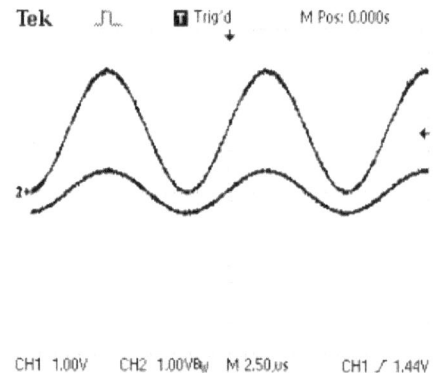

3. Set vertical channel 2 to 200mV/DIV. Measure and record the peak-to-peak voltage at node N3.

    VAC_{N3}: _____

4. Set channel 2 to 500mV/div and the coupling to DC. Connect channel 2 to node N2. Determine the average voltage, V_{AVE}, of the sine wave by algebraically adding the value of its positive and negative peak and dividing by 2.

    V_{AVE}: _____

5. Set vertical channel 2 to 200mV/div and change the coupling to AC. Measure and record the peak-to-peak amplitude of the voltage at node N2.

    VAC_{N2}: _____

72

6. Measure the DC voltages at nodes N1, N2, and N3 with a DMM. Record below:

VDC$_{N1:}$ _____ VDC$_{N2}$: _____ VDC$_{N3}$: _____

Analysis

1. Show that the results of part A can be theoretically calculated using the superposition principle. Model the source as a 1.5V DC source in series with a 3V peak-to-peak AC source. Calculate the percent difference between the measured and calculated values of VDC$_{N2}$ and VAC$_{N2}$.

2. Show that the results of procedure part B can be theoretically calculated using the superposition principle. Model the source as a 1.5V DC source in series with a 3V peak-to-peak AC source. Hint: Consider the capacitor an open circuit for DC and a short (closed) circuit for AC. Calculate the percent difference between the measured and calculated values of: VAC$_{N2}$, VAC$_{N3}$, VDC$_{N2}$, VDC$_{N3}$ and V$_{AVE}$.

Note that: $X_C = \dfrac{1}{2\pi f C}$, XC = ∞ at f = 0 (DC), and X$_C$ = 16Ω at f = 100kHz.

3. Simulate the circuit of part B and compare the results to your measurements and calculations.

Example LTspice Simulation Results for Part B

V1 was set by right clicking on it, setting "Functions" to SINE, DC offset to 1.5, Amplitude to 1.5, Frequency to 100k, and Cycles to 2.

Simulation was set to "Transient", with Stop Time of 20u and Maximum Timestep of .1u.

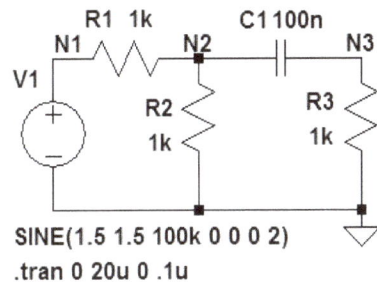

SINE(1.5 1.5 100k 0 0 0 2)
.tran 0 20u 0 .1u

After running the simulation, the graph on the right was obtained by moving the mouse to each node and clicking when the probe appears.

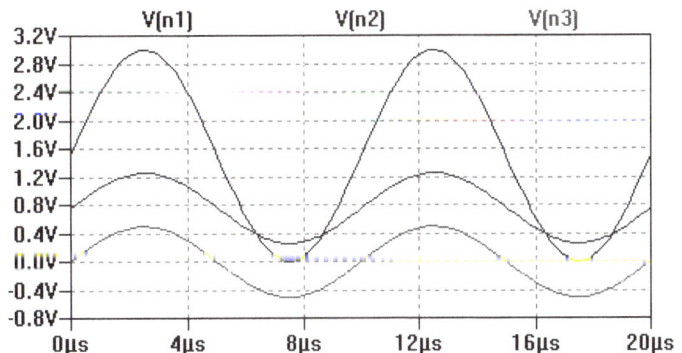

Experiment 22: Working with Phasors

A sinusoidal voltage has a value that is a function of time and may be generally expressed as:

$$v = V\sin(\omega t + \theta) \quad \text{or} \quad v = V\cos(\omega t + \theta).$$

The angle "θ" is usually expressed in degrees. However, the angle "ωt" has units of radians. To evaluate the argument of the sine or cosine, convert angle "ωt" to degrees, then add "θ".

Figure 1 below shows phasor V_a rotating counter clockwise at angular frequency ω. The value of V_a at time t is its vertical component at time t, and is equal to v_a. Figure 1 shows that v_a equals 0 when t is zero. It reaches a maximum when t equals 0.25mSec.

Figure 2 below shows phasor V_a rotating counter clockwise at an angular frequency ω. The value of V_a at time t is its vertical component at time t, which is a cosine function of t and is equal to v_a. Figure 2 shows v_a maximum when t is zero, and zero when t equals 0.25mSec.

Observations:

Period T of V_a=1.0mS.
Frequency f =1kHz.
Frequency ω=6283r/S.

Phasors in polar form for both figure 1 and figure 2:

$$V_a = |V_a| \angle 0^{\circ}$$

$$V_b = |V_b| \angle \theta^{\circ}$$

The only difference between figure 1 and figure 2 is the reference time t = 0. Phasors for figure 1 and figure 2 are identical.

Some textbooks choose the sine function as reference and others use the cosine function to describe sinusoids. As is shown here, the phasor description is identical.

Some textbooks express the magnitude of a phasor in RMS units rather than peak units.

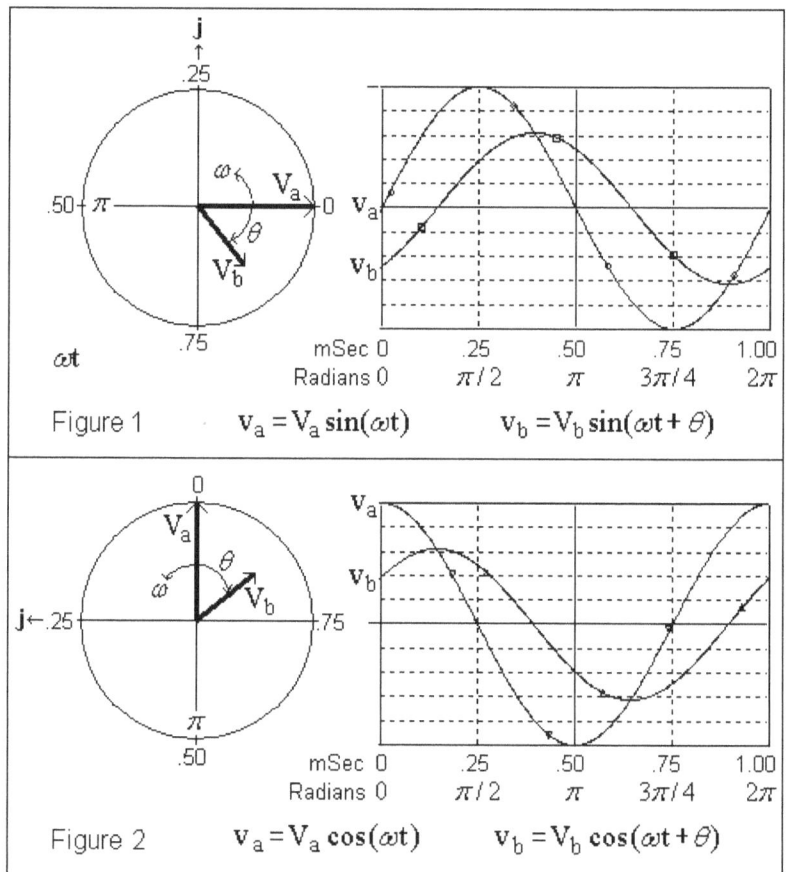

Figure 1 $v_a = V_a \sin(\omega t)$ $v_b = V_b \sin(\omega t + \theta)$

Figure 2 $v_a = V_a \cos(\omega t)$ $v_b = V_b \cos(\omega t + \theta)$

Observe the textbook's conventions when working textbook problems.

It is important to be able to recognize the convention used in contexts outside of your textbook. It is a good idea to include the units when expressing results, for example:

$V_a = 5\angle 30^{0}$ volts peak, $V_a = 3.53\angle 30^{0}$ volts rms, $I_a = 4.6\angle 40^{0}$ mA peak-to-peak.

74

Magnitudes and Angles in LTspice (also in PSpice and other simulation programs)

In LTspice, the magnitude of a sinusoidal source is expressed in peak units. See the example below where the source amplitude is set to 5V. Transient analysis displays the magnitudes of the sinusoids as 5V peak, or 10V peak-to-peak. You can set the phase angle of the source by left clicking on it. This opens the property editor to specify the phase angle at = 0. The phase angle of V1 was set to 0 degrees in the top simulation below and to 90 degrees in the bottom simulation. Refer to the command line for SINE below the schematics.

The first cycle of the waveforms at nodes n001 and n002 differ from the second cycle. It usually takes a few cycles for the waveforms to reach steady state (where each cycle is the same.)

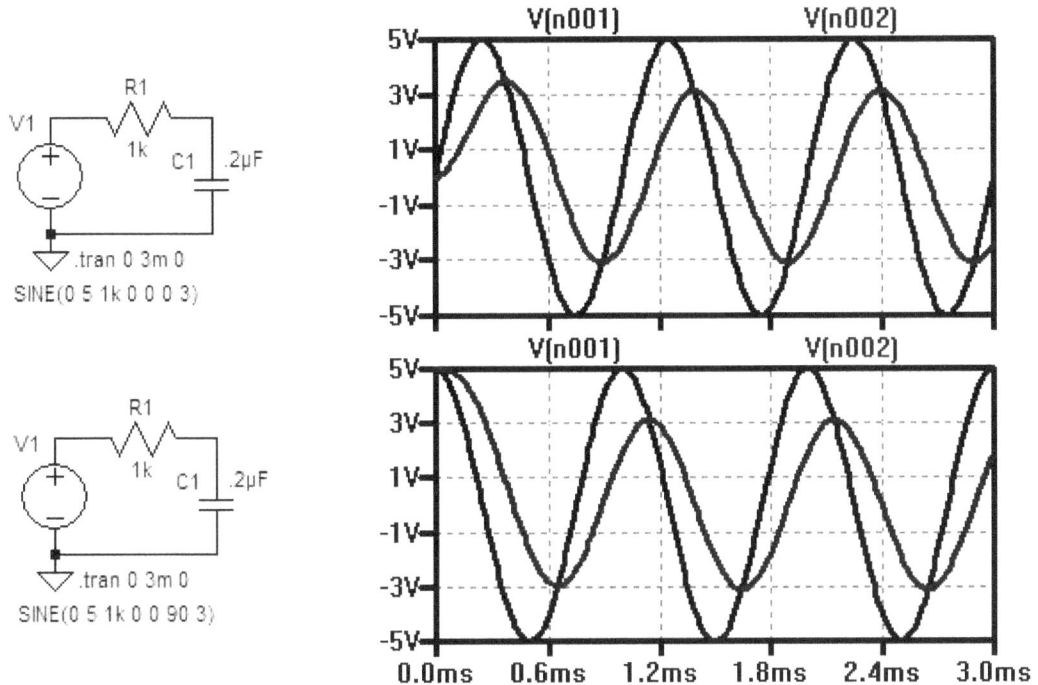

Exercise Suggestions

1. Simulate a homework problem from your textbook involving AC peak-to-peak voltages.

2. Simulate a homework problem from your textbook involving AC rms voltages.

3. Instructor may assign a simulation exercise.

Experiment 23: AC Measurements / Series RC Circuit

Introduction

This lab experiment involves the measurement of the amplitude and phase angle of voltages in a series RC circuit. A "steady state" sinusoidal voltage is applied to the circuit. Steady state refers to a signal whose frequency, amplitude, and phase does not vary with time.

Capacitance has two important properties in circuit analysis. One property is that the voltage developed across a capacitor by a steady state sinusoidal current will be lagging the current by 90 degrees. The other property is called "capacitive reactance", X_C. It is inversely proportional to the frequency of the applied sinusoid.

$X_C = \dfrac{1}{2\pi f C}$ Ohms. f is the frequency in hertz. C is the capacitance in farads.

A series RC circuit provides an opposition to current flow called "impedance". Since the elements are in series the impedance, Z, is given by: $Z = R - jX_C$ where j is the square root of one.

The current, I, can be calculated by Ohm's law: $I = V_S/Z$.

In a series circuit, I is the same in each element. $V_S = V_R + V_C$, and $IZ = IR + I(-jX_C)$.

Procedure

Equipment and Parts
Function Generator, Oscilloscope, and Breadboard. C = 100nF, 5% capacitor, R = 1000Ω, ¼ watt, 5%.

The voltage V_R is in phase with the circuit current in figures A and B below. The circuit of figure B below is used to measure V_C. The resistor and capacitor swap positions to keep the oscilloscope and function generator grounds together. Figure C shows the phase relationship of the circuit voltages with respect to the source, V_S.

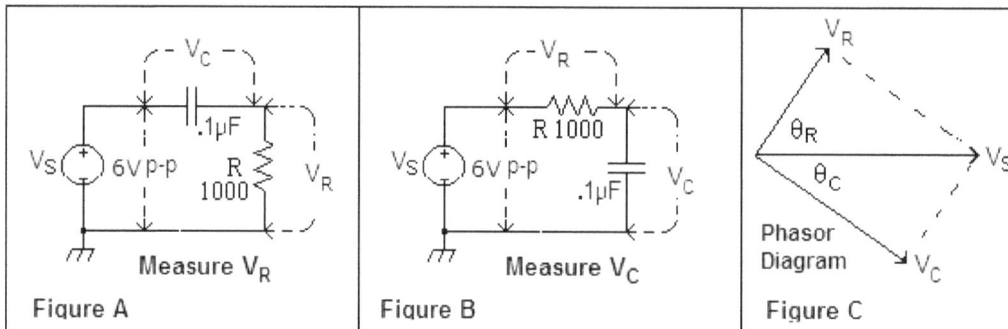

Figure A — Measure V_R

Figure B — Measure V_C

Figure C — Phasor Diagram

1.   Measure and record the value of R and C (C if possible, if not use labeled value).

R_____   C_____

2. Connect the circuit of figure A. Connect oscilloscope channel 1 to measure V_S and channel 2 to measure V_R. Set channel 1 and channel 2 to 1.00V/DIV. Set the time base to 100µS/DIV. Trigger on channel 1. Set the generator to produce a 1000Hz, 6V peak-to-peak sine wave with no offset.

3. Adjust the oscilloscope for a display similar to Figure D below. V_S starts at 0V with a positive slope at the left side of the screen. The "MEASURE" feature is used to measure the peak-to-peak voltages. The cursors are used to measure the time each trace crosses 0V with a negative slope. Record below:

   V_R: _____volts peak-to-peak

4. Note that the period of the sinusoid is exactly 1000µS. V_R crosses zero 160µS before V_S. Therefore V_R leads V_S. The angle can be calculated by the proportion:

$$\frac{160\mu S}{1000\mu S} = \frac{\theta_R}{360^0} \text{ or}: \theta_R = \left(\frac{160\mu S}{1000\mu S}\right)360^0 = 57.6^0.$$

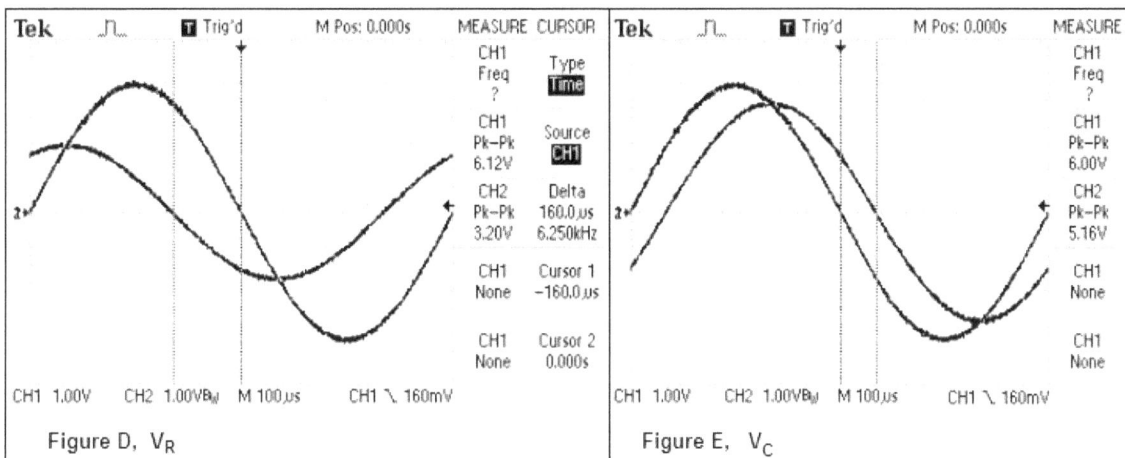

Figure D, V_R

Figure E, V_C

Measure and record the value of θ_R and record below:

   θ_R: _____degrees

5. Connect the circuit of Figure B on page 19 (swap the positions of the resistor and capacitor). Connect oscilloscope channel 1 to measure V_S and channel 2 to measure V_C. Set channel 1 and channel 2 to 1.00V/DIV. Set the time base to 100 µS/DIV and trigger on channel 1. Set the function generator to produce a 1000Hz, 6 V, peak-to-peak sine wave with no offset.

6. Adjust the oscilloscope for a display similar to figure E above. Record below:

   V_C: _____volts peak-to-peak

7. Determine the phase angle, θ_C, and record below:

   θ_C: _____degrees

In figure E, V_C crosses zero 89µS after V_S. Therefore V_C lags V_S.

$$\theta_C = \left(\frac{-89\,\mu S}{1000\,\mu S} \right) 360^0 = -32.1^0.$$

Analysis

1. Calculate the theoretical value (magnitude and phase angle) of the current, I, in the series RC circuit in rms amps. Remember that measurements were taken as peak-to-peak values and must be converted into rms to get results in rms.

$$V_{rms} = \frac{V_{peak}}{\sqrt{2}} = \frac{V_{peak-to-peak}}{2\sqrt{2}}$$

Calculate the percent error between the measured and calculated magnitudes of I (use Ohm's law and the voltage measured across R to calculate the measured value of I).

2. Calculate the theoretical values (magnitude and phase angle) of the voltages, V_R and V_C, in the series RC circuit. Calculate the percent difference between the measured and calculated magnitudes of V_R and V_C. Calculate the absolute difference between their phase angles.

	A	B	C	D	E	F	G	H	I
1	**Magnitudes**	Theoretical	Measured	% Error		**Angles**	Theoretical	Measured	Abs. Error
2	$|V_R|$					$\sphericalangle V_R$			
3	$|V_c|$					$\sphericalangle V_C$			
4	$|I|$					$\sphericalangle I$			

3. Simulate the series RC circuit to obtain the magnitude and phase angle of V_R and V_C. See example below.

LTspice Example

Connect the circuit as shown on the right.

Right click on V1and set "AC amplitude" to 6 under "Small signal AC analysis" in the dialog box that opens.

Labeling the nodes (N1, N2, and N0) makes it easier to interpret the results.

Click on Simulate in the main menu bar.

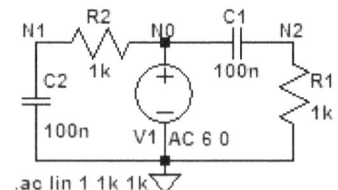

Select *Edit Simulation Cmd*, to open the dialog box shown on the right.

Select *AC Analysis* and enter values as shown. Note that this is for one frequency.

You can specify the simulation to be done at more than one frequency, if desired.

Run the simulation. A window will open listing the simulation results.

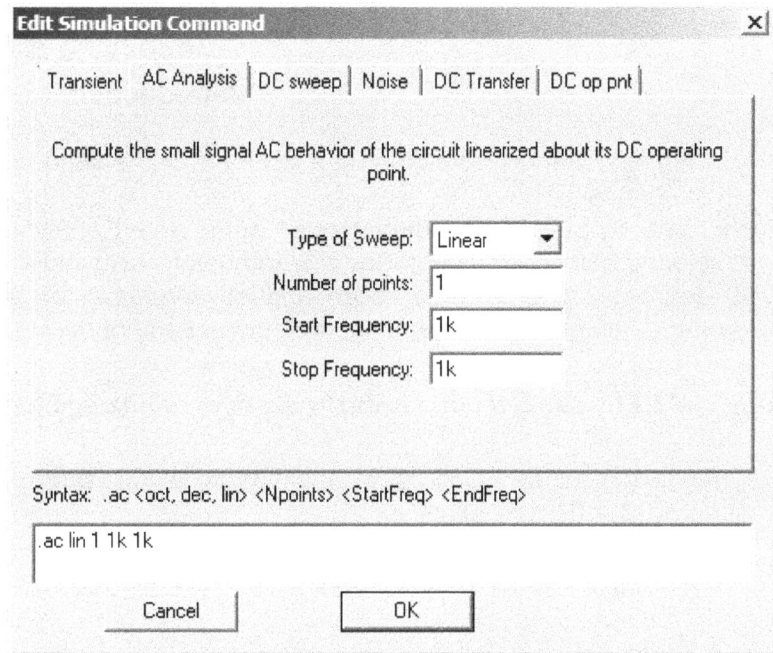

```
         --- AC Analysis ---

frequency:     1000        Hz
V(n2):        mag:    3.19211 phase:     57.8581°          voltage
V(n0):        mag:          6 phase: -4.24074e-015°        voltage
V(n1):        mag:     5.0804 phase:    -32.1419°          voltage
```

History Note:

James Clerk Maxwell (13 June 1831 – 5 November 1879) was a Scottish physicist and mathematician. His most prominent achievement was formulating classical electromagnetic theory. This united all previously unrelated observations, experiments and equations of electricity, magnetism and even optics into a consistent theory. Maxwell's equations demonstrated that electricity, magnetism and even light are all manifestations of the same phenomenon, namely the electromagnetic field. Subsequently, all other classic laws or equations of these disciplines became simplified cases of Maxwell's equations.

From Wikipedia, the free encyclopedia

79

Experiment 24: AC Measurements / Series RL Circuit

Introduction

This lab exercises involves the measurement of the voltages in a series RL circuit.

Inductance has two important properties in circuit analysis. One property is that the voltage developed across an inductor by a steady state sinusoidal current will be leading the current by 90 degrees. The other is called "inductive reactance", X_L. It is directly proportional to the frequency of the applied sinusoid and has units of ohm's.

$X_L = 2\pi f L$ ohms, where f is the frequency in hertz and L is the inductance in henries.

Refer to the series RL circuit at the bottom of this page. The impedance, Z, of the circuit is given by: $Z = (R + R_W) + jX_L$.

The current, I, can be calculated by Ohm's law: $I = V_S/Z$.

In a series circuit, I is the same in each element. Note that the voltage across the inductor in the diagrams below can't be measured directly. $V_X = I(R_W + jX_L)$.

Procedure

Equipment and Parts
Function Generator, Oscilloscope, and Breadboard.
L = 100mH, 5% inductor, R = 1000Ω, ¼ watt, 5%.

At low frequencies an inductor may be approximated by an inductance in series with a resistance. The resistance is mainly due to the wire winding of the inductor. This resistance does increase with frequency, but it is approximately constant for frequencies used in the experiments in this manual (below 10kHz.).

Consider Figure A below. The voltage V_R is in phase with the circuit current. The circuit of figure B below is used to measure V_X. The resistor and inductor swap positions to keep the oscilloscope and function generator grounds together.

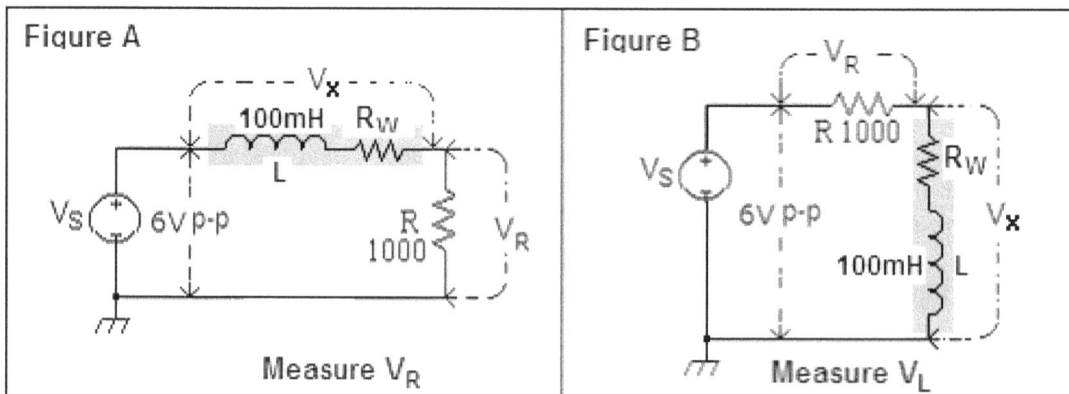

Figure A — Measure V_R

Figure B — Measure V_L

1. Measure and record the value of R, R_W, and L (L if possible, if not, use labeled value).

   R: _____ R_W: _____   L: _____

2. Connect the circuit of figure A. Connect oscilloscope channel 1 to measure V_S and channel 2 to measure V_R. Set Channel 1 and channel 2 to 1V/DIV. Set the time base to 100µS/DIV and Trigger to channel 1. Set the function generator to produce a 1000Hz, 6V peak-to-peak sine wave with no offset as measured by the oscilloscope channel 1.

3. Measure and record the peak-to-peak voltage, V_R: _____.

4. Measure and record the value of θ_R: _____.

5. Connect the circuit of figure B. (swap the positions of the resistor and inductor). Connect oscilloscope channel 1 to measure V_S and channel 2 to measure V_X.

   Set Channel 1 and channel 2 to 1V/DIV. Set the time base to 100µS/DIV and trigger on channel 1. Set the function generator to produce a 1000Hz, 6V peak-to-peak sine wave with no offset as measured by the oscilloscope channel 1.

6. Measure and record the peak-to-peak voltage, V_X: _____.

7. Measure and record the value of θ_X: _____

8. Optional: Repeat this exercise at another frequency, such as 500 Hz or 2000 Hz.

Analysis

1. Calculate the theoretical rms value (magnitude and phase angle) of the current, I, in the series RL circuit. Calculate the percent error between the measured and calculated values of I's magnitude and the absolute error between the phase angles (use the voltage measured across R). Note that you need to include R_W in the calculations and that measurements must be converted to rms to obtain results in rms.

2. Calculate the theoretical values (magnitudes and phase angles) of the voltages, V_R and V_X, in the series RL circuit. Calculate the percent difference between the measured and calculated magnitudes of V_R and V_X and the absolute errors between the phase angles.

	A	B	C	D	E	F	G	H	I
1	Magnitudes (@ 1kHz)	Theoretical	Measured	% Error		Angles (@ 1kHz)	Theoretical	Measured	Abs. Error
2	$|I|$					$\angle I$			
3	$|V_R|$					$\angle V_R$			
4	$|V_X|$					$\angle V_X$			

3. Simulate the circuit using PSpice or LTspice to obtain the magnitude and phase angle of the voltages V_R and V_X. Compare your results to your calculations. They should be in almost exact agreement. Express the percent difference between your measurements and calculations.

Experiment 25: Series-Parallel AC Circuit Measurements

Introduction

Refer to the circuit in the procedure below. The impedance of the circuit in this lab is given by:

$Z_T = (Rw + j\omega L1) + Z_P$, where:

$$Z_p = \frac{-jR1\left(\frac{1}{\omega C1}\right)}{R1 - j\left(\frac{1}{\omega C1}\right)}$$

Rw is the resistance of the inductor. Voltage at node N2 can be calculated using Ohm's law. First calculate the circuit current, I: $I = Vs/Z_T$. Then calculate the voltage developed by that current across the impedance of the parallel combination of C1 and R1. $V_{N2} = I \cdot Z_P$.

Procedure

Equipment and Parts
Function Generator, Oscilloscope with 10X probes, and Breadboard Resistor: 4.7K, ¼ watt, 5%, Inductor: 100mH, 5%, Capacitor: 0.1µF, 5%.

1. Measure and record the values of the components, including the resistance of the inductor, Rw. Connect the circuit. Connect oscilloscope channel 1 to N1 and channel 2 to N2.

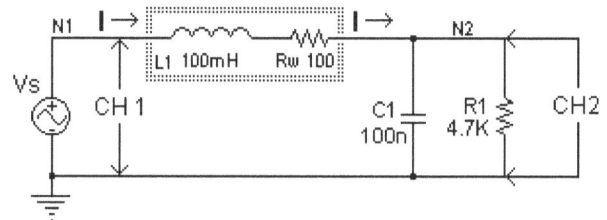

   R1: _____   Rw: _____

   L: _____   C: _____

2. Set the generator to produce a 3.0V peak-to-peak, 800Hz, sine wave at node N1. Measure and record the peak-to-peak magnitude and phase angle of the voltage at node N2 (ch2). Record your results into a spreadsheet table as shown below below.

	A	B	C	D	E	F	G	H	I
1	**Magnitudes**								
2	Frequency	Meas. $\|V_{N2}\|$ (V)	Theor. $\|V_{N2}\|$ (V)	% Error		Theor. $\|Z_T\|$ (Ω)	Meas. $\|V_{N2}\|$ (V)	Theor. $\|V_{N2}\|$ (V)	% Error
3	800								
4	1600								
5	3200								

K	L	M	N	O	P	Q	R
Angles							
Frequency	Theor. $\angle V_R$ (°)	Meas. $\angle V_R$ (°)	Abs. Error (°)		Theor. $\angle I$ (°)	Meas. $\angle I$ (°)	Abs. Error (°)
800							
1600							
3200							

3. Set the function generator to produce a 3.0V peak-to-peak, 1600Hz, sine wave, as measured by channel 1 of the oscilloscope. Repeat step 2.

4. Set the function generator to produce a 3.0V peak-to-peak, 3200Hz, sine wave, as measured by channel 1 of the oscilloscope. Repeat step 3.

Analysis

1. Using measured component values and voltage measurements at node N2, calculate the "measured" circuit current I, at *each* frequency. Enter the results for both magnitude and phase into the table above.

2. Use your scientific calculator to calculate inductive and capacitive reactance at each frequency. Then calculate the impedance of the circuit, Z_T, at each frequency, using the measured values of your components. Enter the results in the table above.

TI-89 example at 1600Hz:

X_{L1}: $2*\pi*1600*.1 = 1005$ ohms. X_{C1}: $= 1/(2*\pi*1600*.1E-6) = 995$ ohms

Z_T: $100+1005i+(-4700*995i)/(4700-995i) = (306.2\angle 9.91)$

To save time and to avoid possible error using the TI-89, you can enter numbers into an equation at multiple places as variables (x and y are used in the example below) and have the calculator evaluate the expression at different values for those variables. This x and y only need to be altered in one place for each new set of x and y values.

$3+5x-2y+x^2/y|x=1$ and $y=2$ This equation yields the result of 4.5
$3+5x-2y+x^2/y|x=2$ and $y=5$ This equation yields the result of 3.8

3. Use Ohm's law, your measured component values, and the source voltage to calculate the theoretical circuit current, I, at each frequency. Enter the results for magnitude and phase in the table.

I: $3/(306.2\angle 9.91) = (9.80\angle-9.91)$.

4. Simulate the circuit. Compare the simulated results to your calculations. They should agree exactly. Express the percent difference between your calculated and measured current and between your calculated and measured node 2 voltage.

LTspice Example

Connect the circuit below. Be sure to use your measured part values.

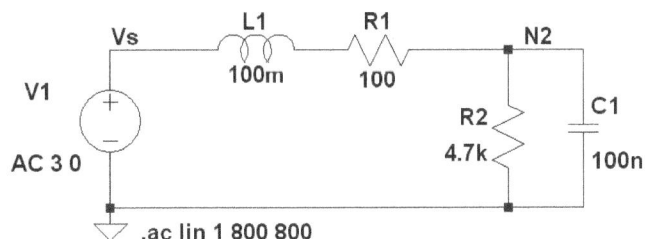

Right click on V1 and set "AC amplitude" to 3 under "Small signal AC analysis" in the dialog box that opens.

Labeling the nodes (Vs and N2) makes it easier to interpret the results. Click on Simulate in the main menu.

Select *Edit Simulation Cmd* to open the dialog box shown on the right.

Select *AC Analysis* and enter values as shown. This is for one frequency. If you specify more than one frequency the output of the simulation will be a graph instead of a list.

You will need to run the simulation three times, once for each frequency.

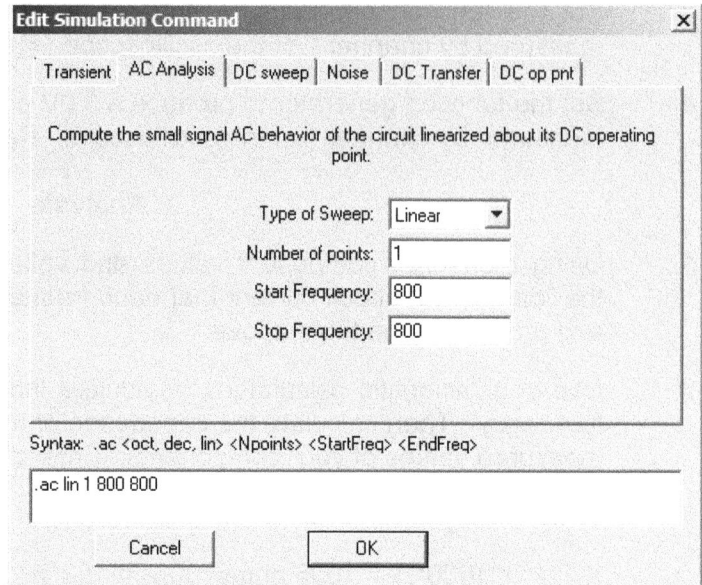

Edit Simulation Command

Transient   AC Analysis | DC sweep | Noise | DC Transfer | DC op pnt |

Compute the small signal AC behavior of the circuit linearized about its DC operating point.

Type of Sweep: Linear ▼

Number of points: 1

Start Frequency: 800

Stop Frequency: 800

Syntax: .ac <oct, dec, lin> <Npoints> <StartFreq> <EndFreq>

.ac lin 1 800 800

Cancel   OK

Run the simulation. A window will open listing the simulation results.

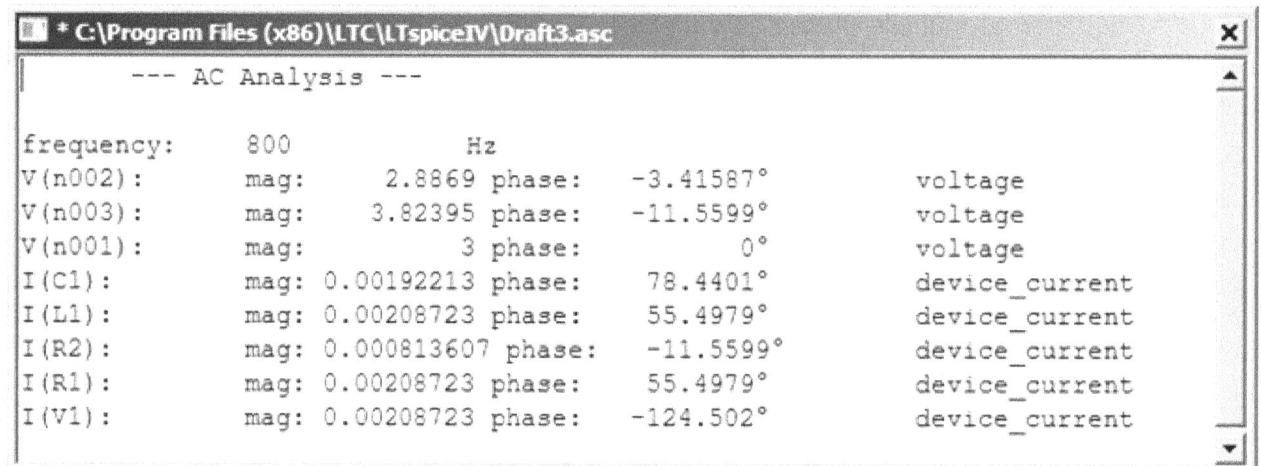

* C:\Program Files (x86)\LTC\LTspiceIV\Draft3.asc

```
      --- AC Analysis ---

frequency:      800            Hz
V(n002):     mag:    2.8869 phase:    -3.41587°      voltage
V(n003):     mag:   3.82395 phase:    -11.5599°      voltage
V(n001):     mag:         3 phase:          0°      voltage
I(C1):       mag: 0.00192213 phase:    78.4401°      device_current
I(L1):       mag: 0.00208723 phase:    55.4979°      device_current
I(R2):       mag: 0.000813607 phase:   -11.5599°      device_current
I(R1):       mag: 0.00208723 phase:    55.4979°      device_current
I(V1):       mag: 0.00208723 phase:   -124.502°      device_current
```

The results for the voltages should agree with your expectations. The currents need to be interpreted according to LTspice conventions. For example, the current in R1 should be in phase with the current in L1. However, note that it may be listed as 180^0 out of phase. The physical orientation of a circuit component in LTspice matters when considering the sign (positive or negative) of the current passing through it. A physical rotation of 180^0 will result in a 180^0 phase shift in the current passing through it.

Run a transient simulation first to insure that the phase angles of the currents will be correct. Place the mouse pointer over each component and observe the current direction arrow. In this circuit the current through L1 and R1 should flow from left to right. The current through R2 and C1 should be flowing down. If the direction through a part does not agree, reverse the direction of that part.

Experiment 26: Capacitive Voltage Dividers

Introduction

Voltage dividers may incorporate inductors and capacitors as well as resistors. Since inductive and capacitive reactance is frequency dependent, the voltage division may also be frequency dependent. In addition, there may be a frequency dependent phase shift.

We will investigate the characteristics of a capacitive voltage divider. The circuit of Figure A below consists of a resistive divider, R1 and R2, and a capacitive divider, C1 and C2. The output voltages of these dividers may be written as:

$$V_{N1} = \frac{R2}{R1+R2} 6V\ p-p,$$

$$V_{N2} = \frac{-j\dfrac{1}{\omega C2}}{-j\dfrac{1}{\omega C1} - j\dfrac{1}{\omega C2}} 6V\ p-p = \frac{C1}{C1+C2} 6V\ p-p.$$

The voltage division is frequency independent. If both dividers divide by the same ratio, that is, $V_{N1} = V_{N2}$, then node N1 may be connected to node N2 without changing the voltage division ratio.

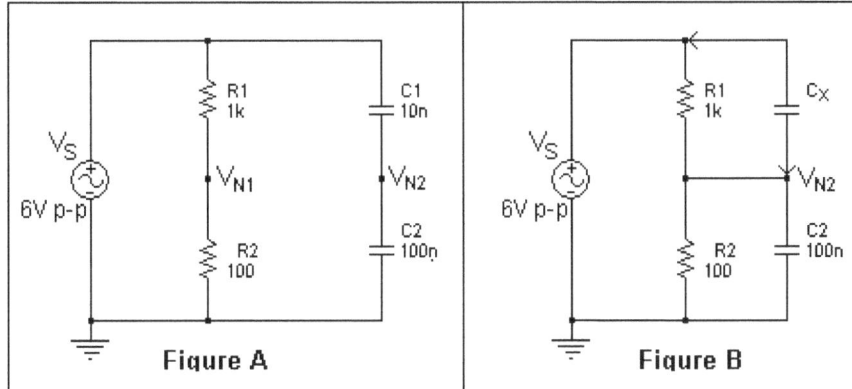

Figure A **Figure B**

An oscilloscope's input can be approximated as a resistance in parallel with a capacitance. The input impedance for a Tektronix TDS1002 is: 1MΩ, ±2% in parallel with 20pF, ±3pF.

In Figure B above, R2 and C2 could represent the oscilloscope input, and R1 and C_X could represent the probe.

Procedure

Equipment and Parts
Oscilloscope, Function Generator, and Breadboard. Resistors: 100 Ω, 5%, 1K Ω, 5%. Capacitors: 10nF, 20nF (or 22nf), 100nF, all 5%.

1.   Measure and record the values of your resistors and (if possible) capacitors.

R1: _____   R2: _____   C2: _____

C_X(10nF):_____ C_X(20nF):_____

2.   Connect the circuit of Figure A on the previous page. Connect oscilloscope channel 1 to measure V_S, and channel 2 to node N1 to measure V_{N1}.

3.   Set the function generator to produce a 6.00V peak-to-peak sine wave with no offset, at a frequency of exactly 8.00kHz. Use the oscilloscope to set the amplitude. This is the setting of the function generator for steps 4 through 7.

4.   Measure and record the amplitude and phase angle of the voltage at node N1 with respect to V_S. Move the oscilloscope channel 2 to node N2. Measure and record the amplitude and phase angle of the voltage at node N2 with respect to V_S.

      V_{N1}: _____ volts p-p θ_{N1}: _____ degrees

      V_{N2}: _____ volts p-p θ_{N2}: _____ degrees

5.   Connect the circuit of Figure B on the previous page without C_X. Connect the oscilloscope channel 1 to measure V_S (6.00V peak-to-peak). Connect the oscilloscope channel 2 to node N2 to measure V_{N2}.

      V_{N2}: _____ volts p-p θ_{N2}: _____ degrees

6.   Connect the circuit of Figure B with C_X = 10nF. Connect the oscilloscope channel 2 to node N2 to measure V_{N2}.

      $V_{N2(10nF)}$: _____volts p-p $\theta_{N2(10nF)}$: _____ degrees

7.   Connect the circuit of Figure B with C_X = 20nF. Connect the oscilloscope channel 2 to node N2 to measure V_{N2}.

      $V_{N2(20nF)}$: _____ volts p-p $\theta_{N2(20nF)}$: _____ degrees

8.   Connect the circuit of Figure B without C_X. Connect the oscilloscope channel 1 to observe V_S. Set the function generator to produce a 6.00V, peak-to-peak, square wave with no offset, at a frequency of exactly 8.00kHz. Use the oscilloscope to set the amplitude. This is the setting of the function generator for steps 9 through 11.

9.   Connect channel 2 to node N2 to observe V_{N2}. Sketch or capture one cycle of the waveform on channel 2.

10.   Connect the circuit of Figure B with C_X = 10nF. Connect the oscilloscope channel 2 to node N2 to observe V_{N2}. Sketch or capture one cycle of the waveform on channel 2.

11.   Connect the circuit of Figure B with C_X = 20nF. Connect the oscilloscope channel 2 to node N2 to observe V_{N2}. Sketch or capture one cycle of the waveform on channel 2.

Analysis

1.   Calculate the theoretical results for procedure steps 5 through 7 and compare the calculations to your measured results.

2. Use the specifications for the vertical input impedance of a Tektronix TDS1002 to design a compensated 10X probe (÷10). The cable connecting the probe to the vertical input adds 15pF to the parallel capacitance. Calculate C_X and R_X in the model on right.

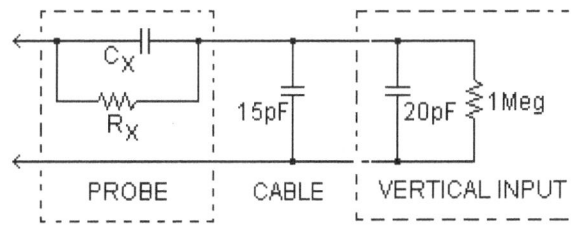

3. (Optional). Compare your results in procedure steps 4 through 7 to simulation. All three cases may be done with one circuit, as suggested below.

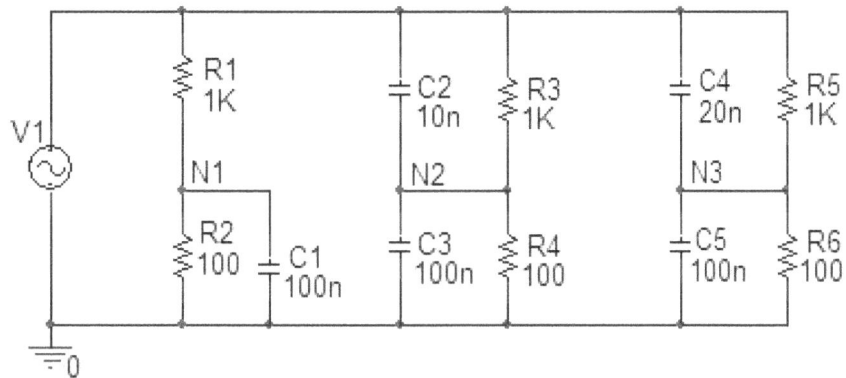

Note that N1 uncompensated, N2 is compensated, and N3 is over compensated.

RESULTS AC ANALYSIS

FREQ	VM(N1)	VP(N1)
8.000E+03	4.961E-01	-2.456E+01

FREQ	VM(N2)	VP(N2)
8.000E+03	5.455E-01	-4.696E-15

FREQ	VM(N3)	VP(N3)
8.000E+03	6.782E-01	1.641E+01

4. Compare your results in procedure steps 9 through 11 to simulation. Replace the AC source with a square wave source (VPULSE). Use transient analysis in the simulation settings.

Experiment 27: Two-Source AC Circuit

Introduction

An AC circuit that is supplied by multiple sources of exactly the same frequency may be analyzed using the phasor method. A "voltage step down" transformer is used in this exercise to supply a circuit with two different voltages of the same frequency. Diagram A below shows a "text book" version of the circuit. Diagram B shows the actual circuit where the transformer provides about 9V peak and 18V peak at the power line frequency of 60Hz. Compare the connections at nodes n1, n2, n3, and n4 in both diagrams.

Throughout this experiment, peak-to-peak values will be used for voltages and current.

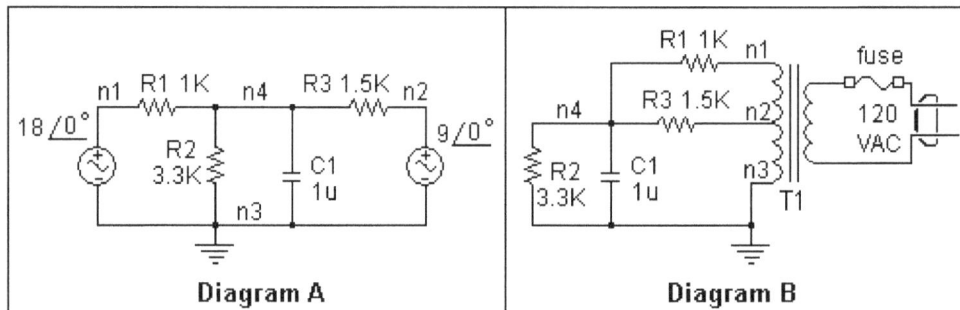

Diagram A   **Diagram B**

Procedure

Equipment and Parts

Transformer: 12.6VAC center tapped, Oscilloscope, and Breadboard.
Resistors: 1K, 1.5K, 3.3K, ¼ watt, 5%. Capacitor: 1µF, non-polarized, 5%.

1.   Measure and record the values of your resistors and capacitor.

   $R1_{1k}$: _____ $R2_{3.3k}$: _____ $R3_{1.5k}$: _____   C1: _____

2.   Connect the circuit in Diagram B above. Power the circuit.

3.   Connect oscilloscope channel 1 to node n1 and channel 2 to node n2.

4.   Measure and record the peak-to-peak values of the voltages at nodes n1 and n2. These voltages should be in phase.

   V1: _____ volts p-p   V2: _____ volts p-p

5.   Connect oscilloscope channel 2 to node n4.

6.   Measure and record the peak-to-peak value of the voltage at node n4 and its phase angle with respect to V1.

   V4: _____ volts p-p   V4: _____ degrees

Analysis

1. Use the measured values of the voltages and the measured values of your components to calculate the voltage and phase angle of the voltage at node n4, V4. Use the node voltage method to write the equation, and a scientific calculator, such as the TI-89, to solve the equation. See example below:

csolve((x4-36)/1000+x4/3300+x4/(-2650i)+(x4-18)/1500=0,{x4})

x4 = V4 = (23.9∠-10.84^0)

2. Simulate the circuit using the measured values of the voltages V1 and V2, and the measured values of your parts. Use transient analysis to generate a display of the voltages at the nodes, similar to the display on the right. Use the cursors to measure the peak-to-peak amplitude and phase angle of the voltage, V4.

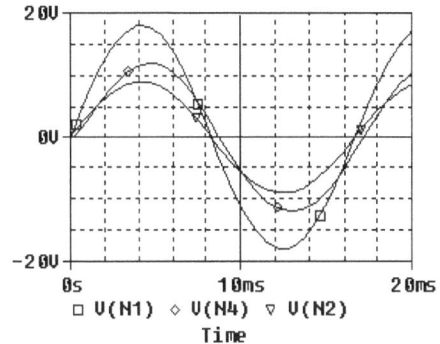

3. Verify that the simulated result for V4 is very close the calculated result.

4. Express the percent error between the measured result and the calculated result for the magnitude of the voltage **V4,** and express the absolute error between the measured and calculated results of the phase angle.

5. Calculate the impedance, **Z2**, of C1 and R2 in parallel. Use the mesh current method to calculate the currents **I1** and **I2**. Refer to the circuit on the right. Use the measured values of your components and the measured values of the voltages **V1** and **V2**.

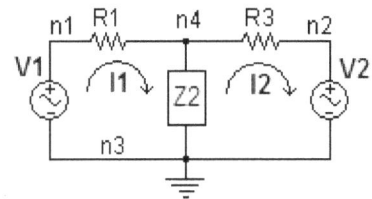

6. Calculate the voltage **V4** using the calculated currents **I1** and **I2**. Compare the result of this calculation with the result of your analysis step 1.

TI-89 Mesh Current Example

Calculate the currents **I1** and **I2** and the voltage, **V4** in the circuit on the right using the mesh current method.

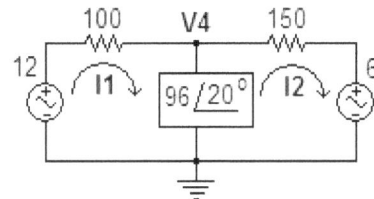

csolve(-12+100*x1+(96∠20)*x1-(96∠20)*x2=0 and
(96∠20)*x2-(96∠20)*x1+150*x2+6=0,{x1,x2})

x1 = **I1** = (61.12∠-7.52) mA   and   x2 = **I2** = (5.35∠94.3) mA

y4=((61.12∠-7.52)-(5.35∠94.3))*(96∠20)   y4 = **V4** = (5.99∠7.67)

Experiment 28: Thevenin's and Norton's Theorems

Introduction

Thevenin's and Norton's Theorems are often used to simplify circuit analysis. They can be used to find the impedance of a complex source. The Thevenin or Norton impedance of any source or circuit is equal to its Thevenin voltage divided by its Norton current. This exercise will also demonstrate the equivalence of a Thevenin circuit to the original circuit.

Procedure

Equipment and Parts
Function Generator, Oscilloscope, and Breadboard   R1 = 1.5K, R2 = 1k, R_L = 1.8K, ¼ watt, ±5%.   C = 100nF, C_L = 47nF, ±5%.

1.   Measure and record the values of R1, R2, R_L, and if possible, C, and C_L.

   R1: _____   R2: _____ R_L: _____

   C: _____ C_L: _____

2.   Connect the circuit of Figure A on the right. Connect the oscilloscope's channel 1 to measure V_S. Trigger on channel 1. Set generator, V_S, to produce an 8V peak-to-peak, 1kHz, sine wave with no offset.

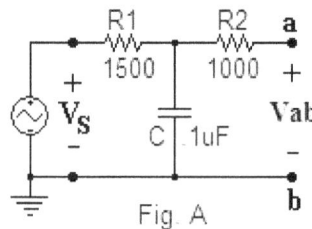

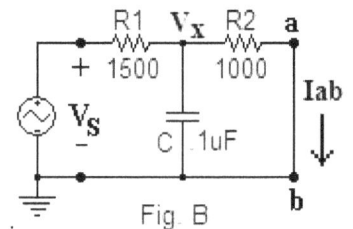

3.   Prepare a spreadsheet table like the one below to record your measurements:

	A	B	C	D	E	F	G
1	Load	Measured Vab - V p-p	Measured Vab - Degrees	Calculated Vab - V p-p	Calculated Vab - Degrees	% Error Vab - V p-p	% Error Degrees
3	Fig. A - None						
4	Fig. A - 1.8K						
5	Fig. A - 47nF						
6	Fig. C - None						
7	Fig. C - 1.8K						
8	Fig. C - 47nF						

4.   Connect the oscilloscope's channel 2 to node a and measure the magnitude and phase angle of the voltage Vab. Record the measurements into cells B3 and C3.

5.   Connect a 1.8K resistor across terminals a and b. Make sure Vs is 8Vpeak-to-peak. Measure the magnitude and phase angle of the voltage Vab and record   into   cells   B4 and C4.

6. Connect a 47nF capacitor across terminals a and b. Make sure Vs is 8V peak-to-peak. Measure the magnitude and phase angle of the voltage, Vab and record into cells B5 and C5.

7. Connect a wire across the terminals a and b as shown in Figure B. Carefully measure and record the magnitude and phase of the voltage, V_X.

$V_{X(mag)}$: _____ $V_{X(phase)}$: _____

8. Connect the circuit in Figure C on the right using the parts used in the circuit of Figure A (disassemble the circuit of Figure A). Set the function generator to the voltage you measured in step 4 (cell B3). This is the circuit's "Thevenin" voltage, V_{TH}.

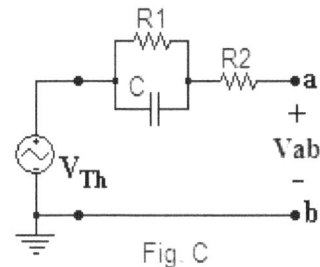

The components are connected in such a way that the impedance between the function generator and terminal **a** is equal to the circuit's Thevenin impedance.

9. Connect the oscilloscope's channel 2 to measure the magnitude and phase angle of the voltage **Vab**. Record the measurements into cells B6 and C6.

10. Connect a 1.8K resistor across terminals **a** and **b**. Make sure that V_{TH} is still equal to the voltage in cell B3. Measure the magnitude and phase angle of the voltage **Vab** and record into cells B7 and C7.

11. Connect a 47nF capacitor across terminals **a** and **b**. Make sure that V_{TH} is still equal to the voltage in cell B3. Measure the magnitude and phase angle of the voltage, **Vab** and record into cells B8 and C8.

Analysis

1. Calculate the theoretical value of the Thevenin voltage, V_{Th}. Compare the result to that measured in step 4. Express the percent difference in the magnitude and phase angle.

2. Calculate the theoretical value of the Norton current, I_N, and compare the result to that measured in step 7. Use V_X to calculate the measured value of I_N. ($I_N = V_X/R2$). Express the percent difference in the magnitude and phase angle.

3. Calculate the theoretical value of the Thevenin impedance using the results of analysis steps 1 and 2 above.

4. Calculate the theoretical value of the Thevenin impedance using the values of the resistors and the reactance of the capacitor. This result should be the same as the results of your calculations in analysis step 3 above

5. Complete the spreadsheet table from step 3 of the procedure. You can use the calculator or simulation for columns D and E and the spreadsheet to calculate columns F and G. Because the function generator Vth angle is 0, add the actual Vth angle to the measured results.

Experiment 29: DC Motor and Generator Basics

Introduction

Permanent magnet DC motors have three basic components, a magnet, an armature, and a commutator. Refer to the diagram on the left below. The motor's armature is a rotating electromagnet. The magnetic force between the armature and the magnetic field of the permanent magnet causes the armature to rotate. The commutator reverses the armature current periodically. This causes the armature's magnetic field to reverse periodically to keep the armature rotating.

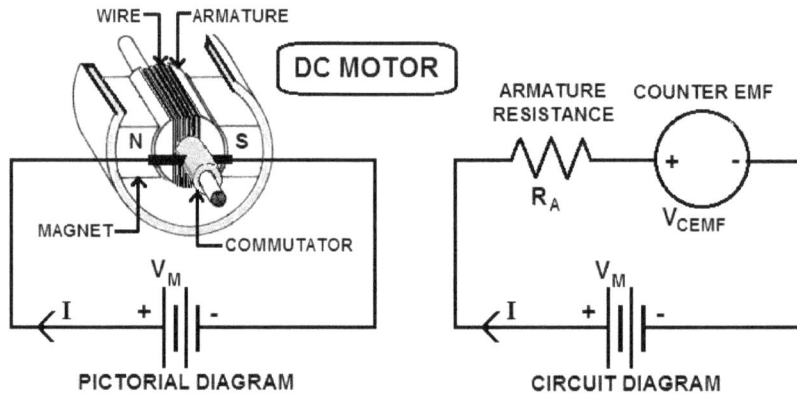

A simplified circuit diagram of the motor is shown above on the right. The resistance of the wire winding of the armature is represented by R_A. As the armature turns a voltage is induced in its winding by the magnetic field of the permanent magnet. This voltage, V_{CEMF}, is called the motor's "counter emf" because its polarity opposes the armature current. Applying Kirchhoff's voltage law to the circuit diagram results in:

$$V_M - I_A R_A - V_{CEMF} = 0 \rightarrow V_{CEMF} = V_M - I_A R_A = K_S S_{RPM}.$$

The counter emf, V_{CEMF} is proportional to the rotational speed of the motor. The proportionality constant is K_S, and the speed in rpm is S_{RPM} in the equation above. K_S has units of volts per rpm.

The armature current is:

$$I_A = \frac{V_M - V_{CEMF}}{R_A} = \frac{V_M - K_S S_{RPM}}{R_A}.$$

If the armature is not rotating, the counter emf is zero and the motor draws its maximum current. When it is rotating, the armature current is reduced by the counter emf. This lab exercise uses a permanent magnet type DC motor to drive a permanent magnet type DC generator.

The rotational speed of the motor-generator set is measured using an "optical interrupter" which generates three pulses per revolution. The frequency of the pulses is measured with the oscilloscope.

Motor and generator speed is usually measured in RPM (rotations per minute). RPM is calculated from the frequency, f, with the equation: RPM = 60f/n, where n is the number of pulses per revolution.

Refer to the diagram of the motor-generator circuit and the terminal labels on the right. The motor power supply is connected between terminals **Vm** and **Gnd**. **Vi** is used to measure the motor current. The magnitude of **Vi** is equal to the magnitude of the motor current. For example, if **Vi** = 450mV, the motor current is 450mA. **Vg** is the output terminal of the generator.

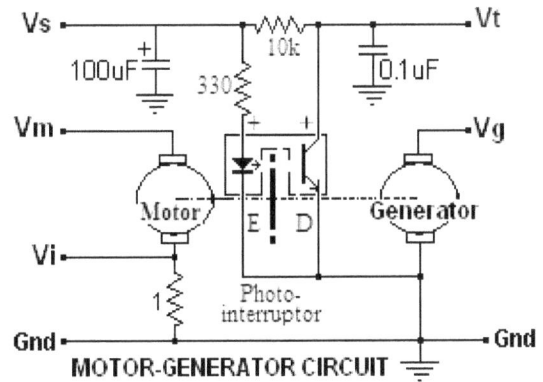

MOTOR-GENERATOR CIRCUIT

Objectives

This exercise demonstrates the basic properties of permanent magnet DC motors and generators. The DC resistance of the motor is measured and used to determine the motor's counter emf as a function of its speed.

Procedure

Equipment and Parts
Power supply: 0 to 6 volts, 1 amp. DMM, Oscilloscope, and Breadboard. Motor-Generator Set (refer to appendix 3). Two 100 Ohm resistors.

Part A: No Load

1.   Measure and record the DC resistance of the motor armature, R_A with an ohmmeter.

   R_A _____

2.   Prepare a spreadsheet table as shown below and record the motor current, generator voltage, and pulse frequency in the table. The last four columns will be calculated.

	A	B	C	D	E	F	G	H
1	**No Load**							
2	Motor Input, Vm (V)	Motor Current (A)	Pulse Width (ms)	Generator No Load Voltage (V)	Calculated Pulse Frequency (Hz)	S_{RPM} (RPM)	V_{CEMF} (V)	K_S (unitless, parts per million)
3	4				–1000/C3	=60*E3/3	=A3-B3*B9	=1000000*G3/F3
4	4.5							
5	5							
6	5.5							
7	6							
9	R_A (Ω)	7.2						

Notice that the formula in cell G3 contains a cell reference of B9. This is a reference to cell to B9 that will not change when the formula is pasted to other cells. The dollar sign locks the reference's row, column, or both, as shown above.

3.   Set the 0 to 6V power supply to 0V. Connect the positive to terminals **Vm** and **Vs** and the negative to ground.

Connect oscilloscope channel 1 to the output of the tachometer, **V$_T$**. Set the trigger to channel 1.

4.   Set the power supply to about 4V. The motor should rotate and a pulse waveform should be observed on channel 1 of the oscilloscope. Set the time base to observe two to three cycles.

5.   Connect the DMM to measure the voltage across the motor: positive red lead to **Vm** and negative black lead to **Vi**. Set the motor voltage to exactly 4.0V.

6.   Measure the frequency of the pulse waveform. Record the frequency, f$_P$, into the table.

Connect the DMM to measure the voltage across the 1Ω resistor: positive red lead to Vi and negative black lead to Gnd. Assuming the resistor is exactly 1Ω, this voltage is equal to the motor current. Be sure the DMM is in a voltage range chosen to give the most accurate reading possible. Record the current into the table.

Move the DMM red lead to **Vg** to measure the generator voltage. Record the generator voltage into the table.

7.   Connect the DMM to measure the voltage across the motor: positive red lead to **Vm** and negative black lead to **Vi**. Set the motor voltage to exactly 5.0V. Repeat step 6 for the motor voltages of 4.5, 5.0, 5.5, and 6.0V.

Optional Part B: Loading Effect

1.   Connect two 100Ω, ¼ watt resistors in parallel to the generator output, Vg. This creates a 50Ω load for the generator.

2.   Repeat part A with the 50Ω load.

	A	B	C	D	E	F	G	H
1	**No Load**							
2	Motor Input, Vm (V)	Motor Current (A)	Pulse Width (ms)	Generator No Load Voltage (V)	Calculated Pulse Frequency (Hz)	S$_{RPM}$ (RPM)	V$_{CEMF}$ (V)	K$_S$ (unitless, parts per million)
3	4				=1000/C3	=60*E3/3	=A3-B3*B9	=1000000*G3/F3
4	4.5							
5	5							
6	5.5							
7	6							
9	R$_A$ (Ω)	7.2						

Analysis, Part A

1. Use the spreadsheet to calculate the motor speed in rpm. $S_{RPM} = 60f_P/n$, where n is the number of holes in the optical interrupter wheel.

2. Use the spreadsheet to calculate the counter emf voltage, V_{CEMF}.

3. Use the spreadsheet to calculate the constant, K_S.

4. Plot the motor speed in rpm as a function of motor voltage.

5. Plot the counter emf, V_{CEMF}, as a function of motor speed in rpm.

6. Plot the motor's constant, K_S, in volts/rpm, as a function of motor speed in rpm.

7. Summarize the results of steps 4, 5, and 6. Is the motor speed proportional to motor voltage? Is the motor's counter emf proportional to the motor speed? Is the motor's constant, K_S, constant?

Analysis: Optional Part B

1. Repeat analysis part A for generator with 50Ω load.

2. Compare step 7 results for part B with results for part A.

Reference: Typical Measurements with No Load

	A	B	C	D	E	F	G
1	Motor Volts	Motor Amps	Fp, Hertz	Gen. Volts	S_{rpm}	V_{cemf}	K_S
3	4.0	0.250	180	2.00	3600.00	2.13	5.90E-04
4	4.5	0.263	205	2.25	4100.00	2.53	6.16E-04
5	5.0	0.275	230	2.50	4600.00	2.94	6.39E-04
6	5.5	0.287	255	2.75	5100.00	3.35	6.56E-04
7	6.0	0.300	280	3.00	5600.00	3.75	6.70E-04

Cell E3: =60*C3/3   Cell F3: =A3-7.5*B3   Cell G3: =F3/E3

Experiment 30: AC Power Basics

Introduction

This lab exercise illustrates the relationship between average power, reactive power, and apparent power.

Fig. 1 on the right shows the phasor diagram for a series RC circuit where the capacitive reactance is equal to the inductive reactance. The resistor and capacitor voltages are equal but 90 degrees out of phase.

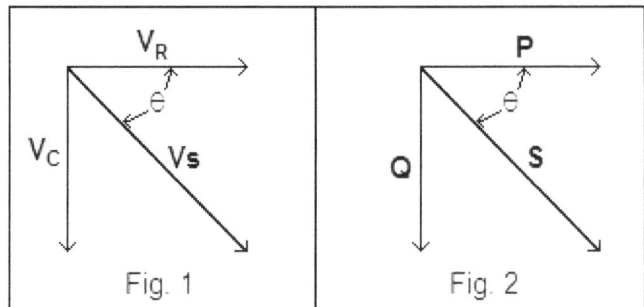

Fig. 1   Fig. 2

Fig. 2 shows the phasor diagram for the circuit power. **P** is the power consumed by the resistor, **Q** is the reactive power of the capacitor, and **S** is the apparent power supplied to the circuit. Note that the angles for the component power correspond to the angles for the component voltage. For this correspondence to be true, the apparent power **S** is calculated as the product of the applied voltage times the complex conjugate of the circuit current: $S = V\,I^*$.

This lab exercise investigates the power relationship in a series circuit where the capacitive reactance is approximately equal to the circuit resistance. In this case, the angle θ in the phasor diagrams above is approximately 45^0. Note that:

$V_R = |V_s|\cos(-45^0)$ and $P = |S|\cos(-45^0)$. $V_C = j|V_s|\sin(-45^0)$ and $Q = j|S|\sin(-45^0)$.

Procedure

Equipment and Parts
Function Generator, Oscilloscope, DMM and Breadboard. Resistor: 1000Ω, ¼ watt, 5%. Capacitor: 100nF, 5%.

1. Measure the value of the resistor:

    R: ____ohms   C: _____ nF

2. Connect the circuit in Figure A. **Vs** is the function generator. Connect channel 1 of the oscilloscope to node n1 and channel 2 to node n2.

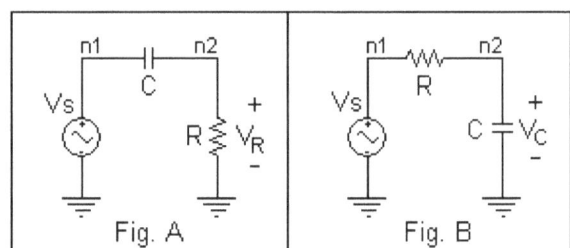

Fig. A   Fig. B

3. Set the function generator to produce a 1600Hz, 10V peak-to-peak, sine wave with no DC offset.

4. Measure and record the peak-to-peak voltage **V_R** at node n2.

    V_R mag: _____volts peak-to-peak.

5. Measure and record the phase of the voltage V_R at node n2 with respect to the voltage **Vs** at node n1.

θ_R: _____degrees. Leading or lagging?

6. Set the DMM to measure AC volts. Measure and record the voltage **Vs** at node n1 and V_R at node n2 with the DMM.

<u>DMM Measurement:</u> **Vs**: _____volts rms. V_R: _____volts rms.

Does the measurement of the magnitude of V_R agree with the oscilloscope measurement of V_R in step 4. Check that:

$$V_R(\text{step } 4) = 2\sqrt{2}\, V_R(\text{step } 6), \text{ within about 2\%.}$$

7. Connect the circuit in Figure B. **Vs** is the function generator. Connect channel 1 of the oscilloscope to node n1 and channel 2 to node n2. Do not change the function generator amplitude and frequency. The amplitude should be 10V peak-to-peak and the frequency should be 1600Hz.

8. Measure and record the peak-to-peak voltage V_C at node n2.

V_C: _____volts peak-to-peak.

9. Measure and record the phase of the voltage V_C at node n2 with respect to the voltage Vs at node n1.

V_C Phase: _____degrees. Leading or lagging?

10. Set the DMM to measure AC volts. Measure and record the voltage Vs at node n1 and the voltage V_C at node n2 with the DMM.

<u>DMM Measurement:</u> **Vs**: _____volts rms. V_C: _____volts rms.

Analysis

1. Note that the purpose of the two circuits was to be able to measure the resistor voltage and the capacitor voltage with respect to the common grounds of the function generator and oscilloscope. Both circuits have the same impedance and therefore the same current. The current can be calculated using the voltage measured across the resistor in step 4 and Ohm's law. **I** = V_R/R.

Calculate and record peak-to-peak magnitude of **I** in mA and the phase angle of **I** in degrees using the voltage and phase angle of VR measured in steps 4 and 5 of the procedure. <u>This is the measured current in because it is calculated from the measured voltage.</u>

I: _____mA peak-to-peak.   **I** Phase: _____ degrees

2. Convert the magnitude of **I** to rms units to obtain the measured current.

I: _____mA rms

97

3. Calculate and record the apparent Power, **S**, supplied to the circuit by the source, **Vs**. Use the measured current from analysis step 2 and measured **Vs** from procedure step 6. Express the result in mVA and in polar form (magnitude and phase angle).

    S: _____mVA   (Measured apparent power)

4. Calculate and record the average Power, P, and the reactive power, Q_C by expressing the value of **S** in step 3 above in rectangular form. These values will be the measured real and reactive power.
    P: _____mW Q_C: _____mVAR.

5. Calculate P and Q_C using values measured by the DMM in procedure steps 6 and 10 and the measured current obtained analysis step 2.

    Compare the result to the analysis 4 result. Although the DMM does not measure phase angle, the results of procedure step 4 (oscilloscope measurements) and procedure step 6 (DMM measurements) should be in close agreement.

6. Calculate the theoretical value of **S** using the voltage source's value of $10V_{p-p}$ (converted to rms) and the total impedance of the circuit, Z_T, which will be calculated from the measured component values. Compare the result to the oscilloscope measurement result of analysis 3 above. Express the percent difference between the theoretical and measured results for the power dissipated by the resistor and for the reactive power of the capacitor.

7. Optional: Simulate the circuit and compare the results to your calculations and measurements.

LTspice Example

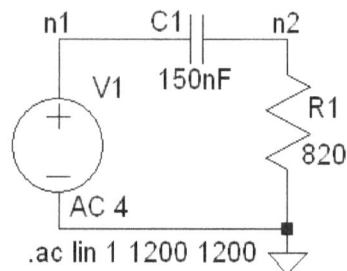

```
--- AC Analysis ---

frequency: 1200 Hz
V(n2):    mag:    2.71996 phase: 47.1572°
V(n1):    mag:          4 phase:       0°
I(R1):    mag: 0.00331702 phase: 47.1572°
```

Source voltage, Vs, is 4.0-volts rms.

$$S = Vs\mathbf{I}^* = 4(.003317\angle - 47.16^0) = 13.27\,mVA = 9.02 - j9.73\,mVA$$

P = 9.02mW dissipated.   Q = 9.73mVARs capacitive.

Experiment 31: 2-Phase Power Distribution

Introduction

Electric power is supplied to a residence by a 240VAC center-tapped transformer. This transformer is supplied by one phase of a three-phase distribution system.

The circuit on the right shows the basic elements involved on the secondary side of the transformer. Power is supplied to the loads R_{L1} and R_{L2} by the transformer. The supply wire has an inductance of L_w and resistance of R_w. Equations for the voltages, V1 and V2, are:

$$V1 = 170\sin(377t)$$

$$V2 = 170\sin(377t + 180) = -V1$$

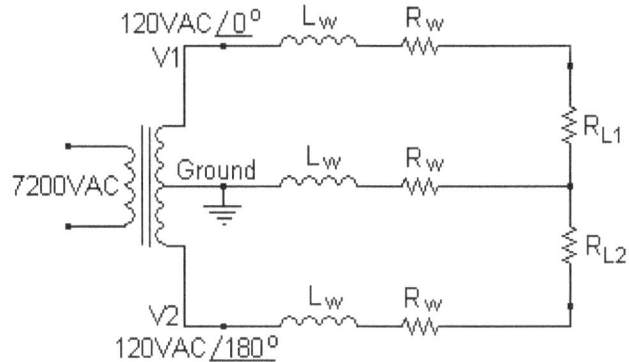

This experiment uses a small low voltage AC transformer to supply power to load resistors. The 100mH inductors simulate line inductance and resistance. For residential service, the transformer is typically mounted on a telephone pole or underground.

The three wires connect to a power meter and an electrical distribution panel in the residence. 240VAC appliances are connected to the two ungrounded wires. They get 240VAC single phase. 120VAC appliances get one phase of the 120VAC.

This lab will demonstrate by measurement and simulation the effect of varying the load resistance and reactance in a power distribution system.

Procedure

Equipment and Parts
Oscilloscope, DMM, and Breadboard.
Transformer: 12VAC, about 1-amp, center-tapped.
L1, L2: 100mH inductor, 50mA minimum, C1: 4.7uF, <u>Non-polarized</u>, 5%
R1: 470Ω, R2: 470Ω, R3: 10Ω, ¼ watt, 5%.

Part A: Basic Measurements

1. Measure and record the resistance of your resistors. Measure and record the resistance, Rw1 and Rw2 of each inductor. If possible, measure the inductance of your inductors (use the lowest frequency on your LCR meter).

    $R1_{470k}$: _____ $R2_{470k}$: _____ $R3_{10}$: _____   C1: _____

    L1: _____   Rw1: _____   L2: _____   Rw2: _____

.

2. Connect the circuit on the right with the transformer unplugged. Lay the circuit out on the breadboard so that it looks like the diagram on the right.

3. Plug in the transformer. Measure and record Va and Vb with respect to ground with the DMM. Set the DMM to read AC rms voltage.

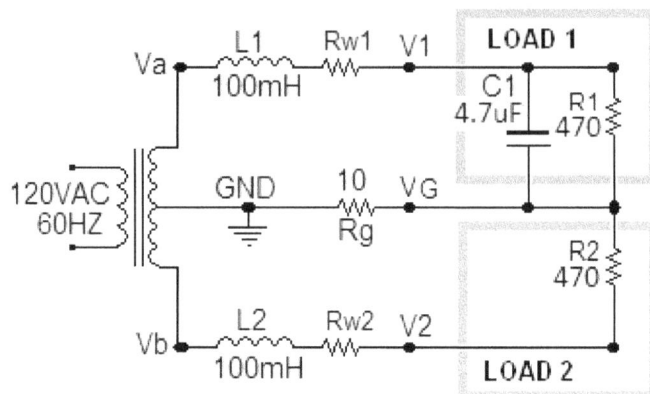

DMM Measurement:

Va: _____V rms.   Vb: _____V rms.

4. Connect channel 1 of the oscilloscope to measure Va with respect to ground (GND). Channel 1 will be the zero degree angle reference.

Measure and record the peak-to-peak amplitude of Va.

Va magnitude: _____V p-p.   Va angle: _____0.0_____degrees.

5. Connect channel 2 to measure Vb with respect to ground (GND). Measure and record Vb. Measure and record the phase angle of Vb (with respect to Va).

Vb magnitude: _____V p-p.   Vb angle: _____degrees.

Verify that Va and Vb p-p magnitudes = $2\sqrt{2}$ times the rms (DMM) magnitudes.

Part B: Oscilloscope Measurements

1. Set up a spreadsheet table like the table below. For each value of load, magnitude and phase measurements are made with the ground connected and disconnected. The 10 ohm resistor, Rg, is used to sample the ground wire current.

Step Row	Gnd	Load	V1 V p-p	V1 Deg.	V2 V p-p	V2 Deg.	VG V p-p	VG Deg.	Ig mA p-p
2	Rg	R1,R2,C2							
3	Rg	R1,R2							
4	Rg	R1							
5	No Rg	R1							
6	No Rg	R1,R2							
7	No Rg	R1,R2,C2							

2. Make and record the measurements indicated in table row 2, (R1, R2, and C1 in circuit).

3. Remove C1 from circuit (R1 and R2 in circuit). Make and record the measurements indicated in table row 3.

4. Remove R2 from circuit. Replace C1 (R1 and C1 in circuit). Make and record the measurements indicated in table row 4.

5. Remove Rg from circuit (R1 and C1 in circuit). Make and record the measurements indicated in table row 5.

6. Remove C1 from the circuit. Reconnect R2 into the circuit (R1 and R2 in circuit, Rg removed). Make and record the measurements indicated in table row 6.

7. Reconnect C1 into the circuit (R1, R2, and C1 in circuit, Rg removed). Make and record the measurements indicated in table row 7.

Analysis: Part B

1. Calculate the magnitude of the current flowing from VG to ground for rows 2, 3, and 4 in the table of procedure part B. Explain why the current is minimum value in row 3.

2. Calculate the voltage, **V1 – VG** and the voltage, **V2 – VG** for table rows 2 and 7. Explain the results considering that row 2 had a ground connection and row 7 did not have a ground connection.

3. Compare the measurement results for table rows 3 and 6. Results should show that a common (ground) wire is not needed if the load is balanced. Relate these results to the results of your calculations in analysis step 2.

4. Write and solve the mesh current equations for the unbalanced load circuit for table row 2 and compare your results to your measurements. Express the percent difference of the magnitudes and the absolute difference of the phase angles.

5. Write and solve the mesh current equations for the unbalanced load circuit for table row 7 and compare your results to your measurements. Express the percent difference of the magnitudes and the absolute difference of the phase angles.

6. Simulate the circuit of procedure step 4 and compare results to your measurements in table row 4 (use your measured component values).

7. Calculate the power factor for the circuit of procedure step 4. Calculate the net power in watts for the resistive circuit elements and in VARS for the reactive circuit elements using the measured magnitudes of the voltages and currents in the table. Show that the phase angle of the circuit current agrees with the result (PF = $\cos^{-1}\theta$).

Note: The instructor may specify the extent of the analysis required for this lab.

LTspice Simulation Example

The circuit on the right is for the simulation that corresponds to table row 2.

Set the analysis to transient to determine the direction of the current through the parts. The current arrows should point to the right through L1 and Rw1, down through R1, R2, and C1, and to the left through Rg, L2, and Rw2.

If the arrows point are in the opposite direction through any part, you will need to reverse the direction of that part.

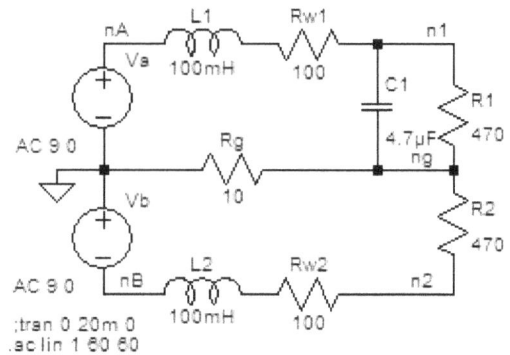

Set the analysis type to AC, linear, 1 point, start frequency to 60, and the stop frequency to 60. The voltage sources, Va and Vb, are set to 9 representing 9V p-p. Set your voltage sources and part values to your measured values for your simulation.

The simulation results are below. The voltage magnitudes are in volts peak-to-peak.

```
        --- AC Analysis ---

frequency:      60              Hz
V(na):          mag:            9 phase:   -3.6223e-015°        voltage
V(nb):          mag:            9 phase:          180°          voltage
V(n001):        mag:      9.38245 phase:     -4.36169°          voltage
V(n002):        mag:      8.98715 phase:      176.203°          voltage
V(n1):          mag:      7.64407 phase:     -12.5118°          voltage
V(n2):          mag:      7.40618 phase:      176.061°          voltage
V(ng):          mag:     0.107568 phase:      73.0275°          voltage
```

History Note:

Sir Charles Wheatstone (6 February 1802 – 19 October 1875), An English scientist and inventor of the English concertina, the stereoscope, and the Playfair cipher (an encryption technique). Wheatstone is best known for his contributions in the development of the Wheatstone bridge

From Wikipedia, the free encyclopedia

102

Experiment 32: Power Factor Compensation / Parallel Circuit

Introduction

This experiment demonstrates power factor compensation of a load using a parallel connected compensation component. The circuit block diagram below shows a voltage source, V_S, with internal resistance, R_S, connected to a load whose apparent power is $S_1 + S_2$.

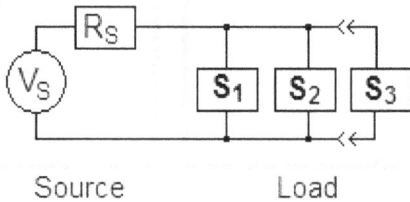

Power factor compensation of the load is accomplished by connecting a component in parallel with the load so that its reactive VARs, S_3, cancel the reactive VARs of $S_1 + S_2$.

That is: $S_{Load} = S_1 + S_2 + S_3 = P$.

In this experiment the source resistance is 1000Ω. The load is a 1000Ω resistor connected in parallel with an inductor. The compensation component will be a capacitor. In addition to power factor compensation, this exercise will demonstrate the frequency sensitivity of the compensated circuit.

Procedure

Equipment and Parts

Function Generator, Oscilloscope, and Breadboard.
Resistors: Two $1K\Omega$, ¼ W, 5%. Inductor: 100mH, 5%. Capacitor: 100nF, 5%

---

1. Measure the value of each resistor, the resistance of the inductor, R_1 and if possible, the values of the inductance, L_1 and capacitance, C.

   R_S: _____ R_1: _____ C: _____ L: _____ Rw: _____

2. Connect the circuit on the right without the capacitor, C. This connection will be referred to as the "uncompensated" circuit.

3. Connect oscilloscope channel 1 to measure **V1** and oscilloscope channel 2 to measure **V2**. Set the trigger to channel 1.

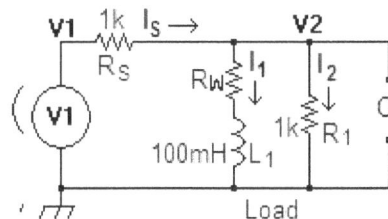

4. Set the function generator, **V1**, to produce a 6V peak-to-peak 1600Hz sine wave with no offset. Set channels 1 and 2 of the oscilloscope to get a display similar to that shown on the next page. Both traces are centered vertically and the reference voltage, **V1**, crosses zero at the horizontal center of the screen.

   The measure feature of the TDS1002 is used to measure the amplitudes of **V1** and **V2**.

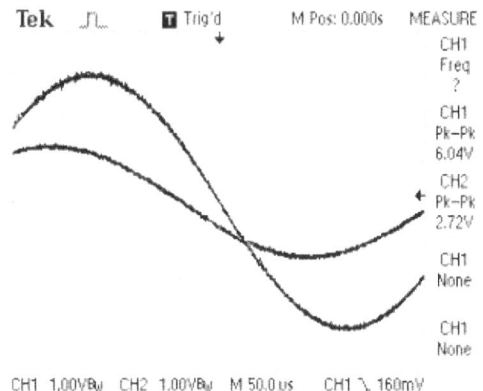

103

V2 crosses zero volts about 44 microseconds before **V1**. It is leading **V1** so its angle is positive:

$$Angle = \frac{44\mu s}{625\mu s} 360 = 25.34°$$

The oscilloscope's time base should be adjusted to measure the time difference between the zero crossings of **V1** and **V2** most accurately.

5.   Set up a spreadsheet as shown below: *Comp means compensated.

	A	B	C	D	E	F	G	H	I	J	K
					V2 Mag	V2 Angle		I Angle	Pout, R1	Pin, Ps	
1	Circuit	Freq (Hz)	T (μs)	t (μs)	(V)	(°)	I (mA)	(°)	(mW)	(mW)	Efficiency
2	Un-Comp	1600	625								
3	Comp	1600	625								
4	Un-Comp	3200	312.5								
5	Comp	3200	312.5								

6.   Measure the peak-to-peak amplitude of **V2** and enter the value into spreadsheet cell E3. Measure the time difference between the zero crossings of **V1** and **V2**. If available, use "cursors" to measure the time difference. Record the results into spreadsheet cell D3. The time difference is positive if **V2** crosses zero before **V1**, and negative if **V2** crosses zero after **V1**.

7.   Connect the 0.1μF capacitor, C1, into the circuit. Check that the generator's output is still 6V peak-to-peak. Measure the magnitude of **V2** and enter the value into spreadsheet cell E4.

Measure the time difference between the zero crossings of **V1** and **V2**. Record the results into spreadsheet cell D4.

8.   Remove the compensating capacitor. Change the generator frequency to 3200Hz. Repeat step 6. Record the results into spreadsheet row 5 (E5 and D5).

9.   Measure the peak-to-peak amplitude of **V2** and enter the value into spreadsheet cell E5. Measure the time difference between the zero crossings of **V1** and **V2**. If available, use "cursors" to measure the time difference. Record the results into spreadsheet cell D5.

10.   Calculate the required value for the compensating capacitance. Connect the capacitance into the circuit, using a parallel combination of capacitors if necessary.

11.   Check that the generator's output is still 6V peak-to-peak. Measure the magnitude of **V2** and enter the value into spreadsheet cell E6.

Measure the time difference between the zero crossings of **V1** and **V2.** Record the results into spreadsheet cell D6.

104

Analysis

1. Calculate the values for columns F through K of the spreadsheet using the measurements in columns C, D, and E. The angle of V2 in column F may be calculated by the spreadsheet. Enter into cell F3: =(D3/C3)*360.

    The current I in column G is calculated: $I = \dfrac{6\angle 0 - V2\angle\theta_2}{R_S}$

    Enter result into spreadsheet cells G and H. Enter current in mA units.

2. Calculate the power P delivered to the load resistor, R1, the power supplied by the source, P1, and the efficiency of the power transfer for the uncompensated and compensated circuits, rows 3, 4, 5, and 6. Record the results in the appropriate cells, I, J, and K. You can input the equations below into cells, I3, J3, and K3 to let the spreadsheet do the calculations.

    P = 1000*V2*V2/(8*R1) mW   Cell I3: =1000*E3*E3/(8*R1) mW

    Ps = 6*I*cos(V2 Deg)/8   Cell J3: =(6*G3*cos(H3*3.14/180))/8

    Ps is in mW because I is in mA. The factor ÷8 converts p-p to rms.

    Note: Enter the value of R1 in ohms. Remember that power must be calculated using rms units. The oscilloscope measurements are in peak-to-peak units. The factor "8" in the equations converts peak-to-peak units to rms units.

3. Simulate the circuit using AC sweep analysis and label nodes N1 and N2. Set the sweep to "Linear" and use one point with start and stop set to 1600, then again to 3200. Use your measured component values. See the example below.

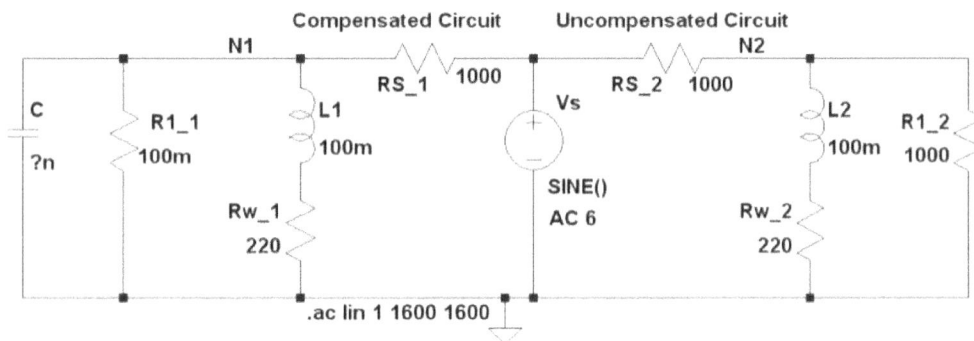

4. Calculate the percent difference between the simulated results and the measured results for the magnitude and phase angle of V2 for the compensated circuits at 1600Hz and 3200Hz.

5. Explain why the compensated circuits deliver more power to the load (R1) than the uncompensated ones.

Experiment 33: Series Compensation and Power Transfer

Introduction

Electrical power is delivered to loads by wire transmission lines. Wire has resistance; therefore power is lost in the wire ($P = I^2 R$). Long transmission lines may also have a significant inductive and capacitive reactance. The voltage developed across the reactance of a transmission line reduces the voltage and power supplied to the load. A capacitor placed in series with a transmission line may be used to compensate for the line's inductance.

Capacitance in series with an inductive transmission can be used to increase the power delivered to the load. Power transmission is improved by making the reactance of the capacitance equal to the inductance of the transmission line. Recall that the net reactance for inductance and capacitance in series is: $X = j(X_L - X_C)$.

Also recall that the maximum power transfer theorem for AC states that maximum power transfer occurs when the load impedance is equal to the complex conjugate of the source impedance.

In this experiment the load is a 330Ω resistor, R2. A 100mH inductor simulates the transmission line inductance and resistance. A compensating capacitor will be placed in series with the inductor. This exercise will show that much more power is delivered to the load by the compensated line.

Procedure

Equipment and Parts
Function Generator, Oscilloscope, and Breadboard
Resistor: 330Ω, ¼ watt, 5%. Inductor: 100mH, 5%. Capacitor: 0.1µF, 5%.

1. Measure the resistance of the resistor, the resistance of the inductor, and if possible, the capacitance of the capacitor and inductance of the inductor.

   R2: _____   C1: _____   L1: _____   R$_W$: _____

2. Connect the circuit below with a jumper wire across the capacitor. This will be the uncompensated circuit <u>without the capacitor</u>, C1.

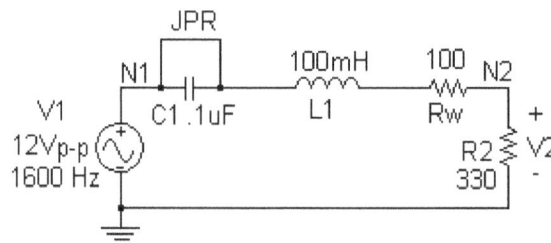

3. Set the function generator to produce a 12.0V peak-to-peak, 1600Hz, sine wave, V1, at node N1, as measured by the oscilloscope channel 1.

4.   Set up a spreadsheet for your measurements as shown below:

	A	B	C	D	E	F	G	H	I	J	K
1	Circuit	Freq	T mS	t mS	V2 Mag	V2 Deg	I mA	I Deg	P mW	Pin mW	
3	Uncomp.	1600	0.625								
4	Comp	1600	0.625								
5	Uncomp	3200	0.3125								
6	Comp	3200	0.3125								

Note: Columns F through J will be calculated by the spreadsheet. Instructions for the calculations are given in the analysis section of this lab exercise.

5.   Connect channel 2 of the oscilloscope to node N2. Measure the peak-to-peak value of the voltage, V2, at node N2 and record in cell E3. Measure the time between the zero crossings of V2 and V1 (in milliseconds). Record the result in cell D3.

6.   Remove the jumper wire across the capacitor. This will be the compensated circuit <u>with the capacitor</u>, C1.

7.   Set the function generator to produce a 12.0V peak-to-peak, 1600Hz, sine wave, V1, at node N1, as measured by the oscilloscope channel 1.

8.   Connect channel 2 of the oscilloscope to node N2. Measure the peak-to-peak value of the voltage V2, at node N2 and record in cell E4. Measure the time between the zero crossings of V2 and V1 and record in cell D4.

9.   Remove the compensating capacitor from the circuit and replace the jumper wire.

10. Set the function generator to produce a 12.0V peak-to-peak, 3200Hz sine wave, V1, at node N1, as measured by the oscilloscope channel 1.

11. Connect channel 2 of the oscilloscope to node N2. Measure the peak-to-peak value of the voltage V2, at node N2 and record in cell E5. Measure the time between the zero crossings of V2 and V1 and record in cell D5.

11. Calculate the required value of the compensating capacitance for 3200 Hz. Replace the jumper with the compensating capacitor, use capacitors in parallel if necessary.

12. Set the function generator to produce a 12.0V peak-to-peak, 3200Hz, sine wave, V1, at node N1, as measured by the oscilloscope channel 1.

13. Connect channel 2 of the oscilloscope to node N2. Measure the peak-to-peak value of the voltage V2, at node N2 and record in cell E6. Measure the time between the zero crossings of V2 and V1 and record in cell D6.

Analysis

1. Enter the following equations into the indicated cells:

   Cell F3: =(D3/C3)*360   Cell G3: =1000*E3/R2   Cell H3: =F3

   Cell I3: =1000*(E3*E3)/(8*R2)   Cell J3: =(12*G3*cos(H3*3.14/180))/8

   Verify that the above equations are correct.

2. The example spreadsheet below was produced using simulation data and the labeled values of capacitance and inductance. Your results should be within about 20% of those in the spreadsheet below.

	A	B	C	D	E	F	G	H	I	J
1	Circuit	Freq	T mS	t mS	V2 Mag	V2 Deg	I mA	I Deg	P mW	Pin mW
3	Uncomp	1600	0.6250	-0.116	3.62	-66.8	10.97	-66.8	4.96	6.49
4	Comp	1600	0.6250	-0.002	9.21	-1.4	27.91	-1.4	32.13	41.85

3. Simulate the circuit using AC analysis. Set the sweep to linear, 1 point, start frequency to 1600 and stop frequency to 1600. Be sure to use your measured component values. See example below:

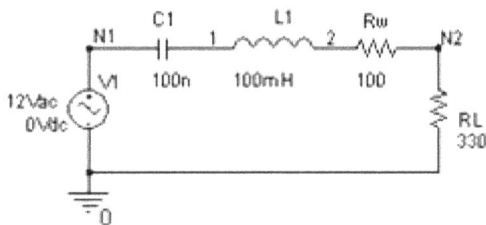

Compensated		
FREQ	VM(N2)	VP(N2)
1.600E+03	9.207E+00	-1.411E+00

Uncompensated		
FREQ	VM(N2)	VP(N2)
1.600E+03	3.622E+00	-6.684E+01

4. Calculate the percent difference between the simulated results and the measured results for the magnitude and phase angle of V2.

5. Explain why the compensated circuit delivers more power to the load (R2) than the uncompensated one at 1600Hz.

6. Explain the results of your measurements at 3200Hz.

	A	B	C	D	E	F	G	H	I	J	K
1	Circuit	Freq.	T mSec	t mSec	V2 Mag	V2 Deg	I mA	I Deg	P mW	Ps mW	
3	Un-comp	1600	0.625								
4	Comp	1600	0.625								
5	Un-comp	3200	0.3125								
6	Comp	3200	0.3125								

108

Experiment 34: Low-Pass Filters

Introduction

Filters are frequency selective networks used to pass some frequencies and attenuate others. A filter is characterized by its "pass-band", "stop-band", and "cutoff frequency". Its cutoff frequency is also called its "half-power frequency" and its "minus 3dB frequency". This is the frequency where the filter's power output is exactly equal to one-half of its power input. This occurs when the magnitude of its output voltage is equal to the magnitude of its input voltage divided by the square root of two.

Consider the low-pass RC filter circuit on the right. The magnitude of its transfer function and its cutoff frequency, f_c, are given by:

$$\left|\frac{V_o}{V_i}\right| = \frac{1}{\sqrt{1 + (2\pi f RC)^2}} \qquad f_c = \frac{1}{2\pi RC} = 338.65\,\text{Hz}.$$

A log-log plot of the frequency response of the low-pass filter is shown on the right. V_i is 10 volts. V_o is 7.07 volts (-3dB) at the cutoff frequency of 340Hz.

V_o decreases by a factor of 10 for each decade increase in frequency beyond the cutoff frequency (-20dB/decade).

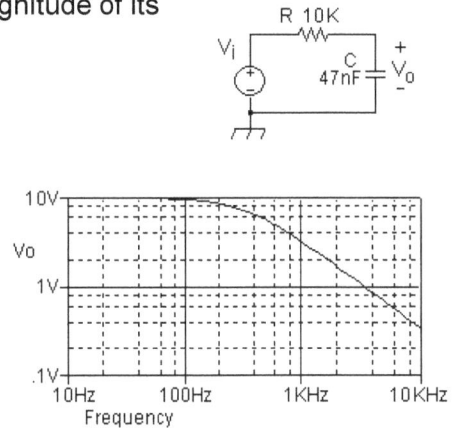

The effect of a load resistance connected across the capacitor can be calculated by finding the Thevenin equivalent circuit external to the capacitor. Replace R in the equations above with the circuit's Thevenin resistance, R_{Th}, and V_i with the circuit's Thevenin voltage, V_{Th}.

Consider the low-pass RL filter circuit on the right. Rw is the inductor's winding resistance. The magnitude of its transfer function and its cutoff frequency are given by:

$$\left|\frac{V_o}{V_i}\right| = \frac{R}{(R + Rw)\sqrt{1 + (2\pi f L/(R + Rw))^2}} \qquad f_c = \frac{R + Rw}{2\pi L}$$

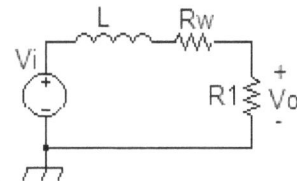

The inductor's wire resistance, Rw, and the filter's output resistance, R, make a voltage divider that reduces the filter's maximum output voltage. Rw also affects the filters frequency response.

Objectives

The frequency response of RC and RL low-pass filters will be investigated, including the effect of load resistance on the response of the RC filter and inductor wire resistance on the response of the RL filter.

Procedure

Equipment and Parts
Function Generator, Oscilloscope with 10X probes, DMM.
R1 = 10K, R2 = 2.2K, RL = 33K, all ¼ watt, 5%.
L = 100mH, C = 47nF, 5%.

Measure and record the value of the parts.

R1: _____   R2: _____   RL _____

L: _____   Rw: _____   C _____

Part A: RC Low-Pass Filter

1. Connect the circuit on the right. Connect oscilloscope channel 1 to the function generator, Vi, and channel 2 to measure Vo.

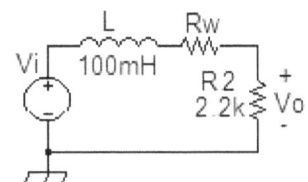

2. Set trigger to channel 1 positive slope. Set the oscilloscope to measure the amplitude and phase angle of Vo with respect to Vi.

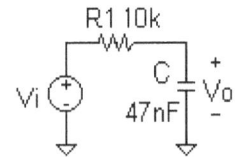

Set the generator to produce a 10Hz, 10 V peak-to-peak, sine wave.

3. Measure the amplitude and phase angle of Vo at 10, 20, 100, 1000, and 10,000Hz. Record the data in the table below.

Measure the frequency and phase angle of Vo at the frequencies where the amplitude of Vo is 8.50, 7.07, and 5.00V peak-to-peak. Record the data in the table on the right.

Frequency Hz	Vo volts p-p	Degrees
10		
20		
100		
	8.50	
	7.07	
	5.00	
1000		
10,000		

Part B: RC Low-Pass Filter with Load Resistor

1. Connect a 33K load resistor across the capacitor of your low pass filter.

2. Set the function generator to produce a 10Hz, 10V peak-to-peak, sine wave.

3. Measure and record the peak-to-peak output voltage, Vo. Vo _____

4. Tune the function generator to the frequency where the peak-to-peak output voltage, Vo, is 0.707 of its value in step 3 above. Record the frequency.

    f_c _____

Part C: RL Low-Pass Filter

1. Connect the circuit on the right. Connect oscilloscope channel 1 to the generator, Vi, and channel 2 to measure Vo.

110

2. Set the oscilloscope to measure the amplitude and phase angle of Vo with respect to Vi, with trigger on channel 1. Set the generator to produce a 100Hz, 10V peak-to-peak, sine wave.

3. Measure and record the amplitude and phase angle of Vo at 100, 200, 1000, 10000, and 100,000Hz.

Measure and record the frequency and phase angle of Vo at the frequencies where the amplitude is 8.50, 7.07, and 5.00 V p-p.

Freq. Hz	Vo volts p-p	Degrees
100		
200		
1000		
	8.50	
	7.07	
	5.00	
10,000		
100,000		

Analysis Part A, Low-Pass RC Filter

1. Plot the frequency response of the filter on semi-log paper. Plot the frequency on the log axis, and the amplitude in dB on the vertical axis ($dB = 20 \log Vo$).

   Optional: Plot the frequency response with a spreadsheet.

2. Calculate the filter's cutoff frequency and the percent difference between the measured and calculated values.

3. Calculate the slope of the filter's attenuation in dB per decade using the amplitudes at 1000Hz and 10,000Hz. Compare result to theoretical value.

Analysis Part B, Low-Pass RC Filter with Load

1. Use Thevenin's Theorem. Find the filter's cutoff frequency and maximum output amplitude.

2. Calculate the percent difference between the measured and calculated values.

Analysis Part C, Low-Pass RL Filter

1. Plot the frequency response of the filter on semi-log paper. Plot the frequency on the log axis and the amplitude in dB on the vertical axis ($dB = 20 \log Vo$).

   Optional: Transfer data to a spreadsheet. Plot the frequency response with the spreadsheet.

2. Calculate the slope of the filter's cutoff frequency and the percent difference between the measured and calculated value.

3. Calculate the filter's attenuation in dB per decade using the amplitudes at 10,000Hz and 100,000Hz. Compare result to theoretical value.

111

Experiment 35: High-Pass Filters

Introduction

Filters are frequency selective networks used to pass some frequencies and attenuate others. A filter is characterized by its "pass-band", "stop-band", and "cutoff frequency". Its cutoff frequency is also called its "half-power frequency" and its "minus 3-dB frequency". This is the frequency where the filter's power output is exactly equal to one-half of its power input. This occurs when the magnitude of its output voltage is equal to the magnitude of its input voltage divided by the square root of two.

Consider the high-pass RC filter circuit on the right. The magnitude of its transfer function and its cutoff frequency, f_c, are given by:

$$\frac{V_o}{V_{in}} = \frac{1}{\sqrt{1 + (1/2\pi f RC)^2}} \qquad f_c = \frac{1}{2\pi RC} = 338.65\,Hz.$$

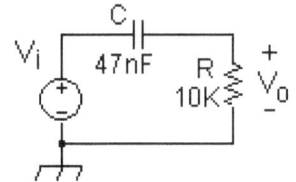

A log-log plot of the frequency response of the filter is shown on the right. Vo is 7.07 volts (-3dB) at the cutoff frequency of 338Hz.

Vo decreases by a factor of 10 for each decade decrease of frequency below the cutoff frequency (-20dB/decade).

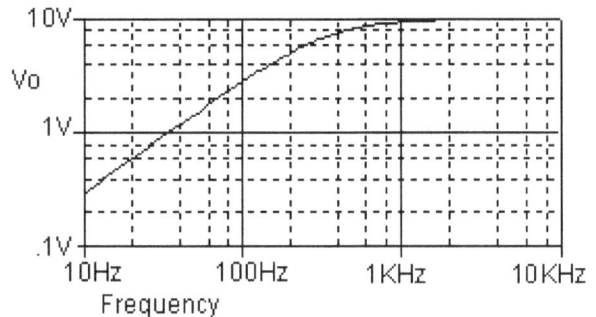

Consider the effect of adding a source resistance, Rs, to a high-pass RC filter circuit as shown on the right. The magnitude of its transfer function and its cutoff frequency, f_c, are given by:

$$\frac{V_o}{V_{in}} = \frac{R/(R + Rs)}{\sqrt{1 + (1/2\pi f (R + Rs)C)^2}} \qquad f_c = \frac{1}{2\pi (R + Rs)C} = 308.0\,Hz.$$

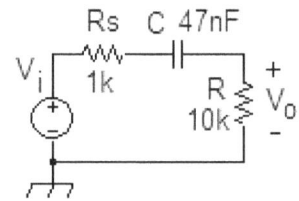

Procedure

Equipment and Parts
Function Generator, Oscilloscope with 10X probes, and Breadboard. R = 10K, Rs = 1K, all ¼ watt, 5%. C = 47nF, 5%.

Measure and record the values of the parts.

R _____   Rs _____   C _____

Part A: High-Pass RC Filter

1. Connect the circuit on the right:

2. Connect channel 1 of the oscilloscope to the function generator, Vi, and channel 2 to measure Vo.

3. Set the trigger to channel 1 and positive slope.

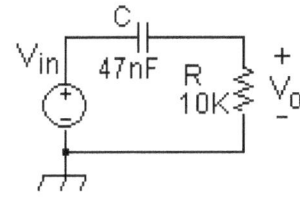

Set the oscilloscope to measure the amplitude of Vo and the phase angle of Vo with respect to Vi.

Set the function generator to produce a 1000 Hz, 10 V peak-to-peak, sine wave (no DC offset).

4. Measure the amplitude and phase angle of Vo at 10, 100, 1000, and 10000Hz. Record the data in the table below.

Measure the frequency and phase angle of Vo at the frequencies where the amplitude of Vo is 5.00, 7.07, and 8.50V peak-to-peak. Record the data in the table below.

Frequency Hz	Vo volts p-p	Degrees
10		
100		
	5.00	
	7.07	
	8.5	
1000		
10,000		

Part B: High-Pass RC Filter with Source Resistance

1. Connect a 1K resistor in series with the generator and your high-pass filter as shown in the diagram on the right.

2. Set the function generator to produce a 10kHz, 10V peak-to-peak, sine wave.

3. Measure and record the peak-to-peak output voltage, Vo. Vo

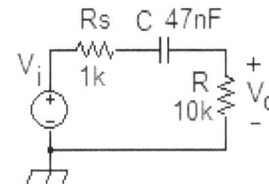

_____

4. Tune the function generator to the frequency where the peak-to-peak output voltage, Vo, is 0.707 of its value in step 3 above. Record the frequency.

    f_c _____

Analysis Part A: High-Pass Filter

1. Plot the frequency response of the filter on semi-log paper. Plot the frequency on the log axis and the amplitude in dB on the vertical axis ($dB = 20 \log Vo$).

2. Calculate the filter's cutoff frequency and the percent difference between the measured and calculated value.

3. Calculate the filter's attenuation in dB per decade using the amplitudes at 100Hz and 10Hz. Compare result to theoretical value.

4. Optional: Simulate the frequency response of the filter.

Analysis Part B: High-Pass Filter with Source Resistance

1. Find filter's cutoff frequency and maximum output amplitude.

2. Calculate the percent difference between the measured and calculated values.

LTspice Example: Part B

Connect the circuit using your measured part values. Set the AC amplitude of the voltage source, V1, to 10 and the AC phase to 0. Set AC Analysis to decade sweep, 10 points per decade, start frequency to 10, and stop frequency to 10k.

By default, the amplitude on the vertical axis will be displayed in decibels. This can be changed by left clicking on the vertical axis, selecting "Manual Limits" selecting linear, and changing the amplitude range to 0 to 10.

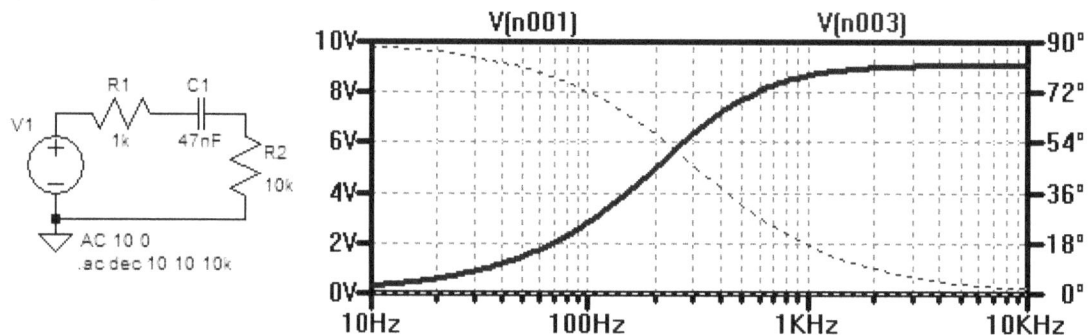

A cursor may be obtained by right clicking on the node label at the top of the graph. Select "1st" in the "Attached Cursor" box. A window will open which displays the magnitude and phase angle as the cursor is moved along the trace.

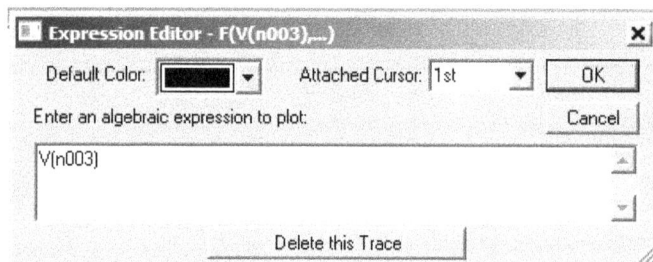

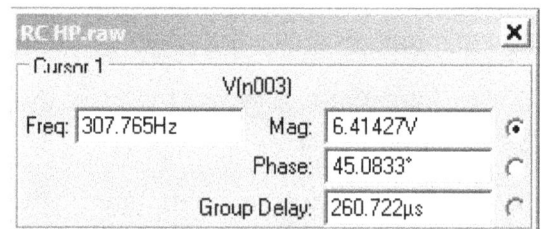

The cutoff frequency is found by moving the cursor to where the amplitude is 0.707 of the band-pass value. Note that the phase angle is 45^0 at the cutoff frequency.

114

Experiment 36: Audio Crossover Network

Introduction

Crossover networks are often used in audio amplifier circuits to separate the low audio frequencies and high frequencies. The crossover frequency for the filter circuit in this lab exercise is about 1600Hz. Frequencies above 1600Hz are passed to the V_{HP} output. Frequencies below 1600Hz are passed to the V_{LP} output.

This crossover network is to be used in an audio amplifier whose overall frequency response is specified to be 10Hz to 20kHz. Part A will evaluate the frequency response of each filter. Part B will evaluate the overall frequency response by summing the output of the two filters.

Procedure

Equipment and Parts
Function generator, Oscilloscope with 10X probes, and Breadboard Resistors: Two 1K, ¼ watt, 5%. Inductor: 100 mH, 5%. Capacitor: 100 nF, 5%. Rw* ≈ 100, ¼ watt, 5%. Select value approximately equal to inductor's resistance.

Part A: Frequency Response

1. Measure and record the DC resistance of the inductor.

   Rw: _____

   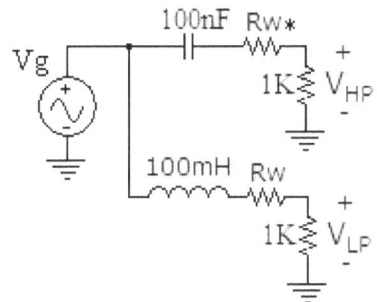

   Connect the circuit. The source, Vg, is the function generator. Rw is the resistance of the inductor. Rw* in the high-pass filter is selected to have approximately the same resistance as the inductor.

2. Connect the function generator, Vg, to the circuit. Connect channel 1 to measure Vg and channel 2 to measure V_{LP}. Set the function generator to produce a 1600Hz, 1V peak-to-peak, sine wave.

   The output of each filter at the cutoff frequency is reduced by the factor: Rw/(Rw + 1k). At the cutoff frequency, the amplitude V_{CF} of V_{HP} and V_{LP} will be given by:

   $$V_{LP}=V_{HP}=V_{CF}=\frac{1000Vg}{\sqrt{2}\,(1000+Rw)}, \ if \ Rw=100, \ V_{CF}=\frac{1000}{\sqrt{2}\,(1100)}=0.643V \ p-p.$$

   Calculate the value of V_{CF} at the cutoff frequency of your filter and record below.

   V_{CF}: _____ V p-p.

3. Connect channel 2 of the oscilloscope to measure V_{LP}. Fine-tune the function generator so that the magnitude of V_{LP} is exactly V_{CF} volts peak-to-peak. Record the resulting frequency as the cutoff frequency of the low pass filter, f_{CLP}.

   f_{CLP}: _____

4. Connect channel 2 of the oscilloscope to measure V_{HP}. Fine-tune the function generator so that the magnitude of V_{HP} is exactly V_{CF} volts peak-to-peak. Record the resulting frequency as the cutoff frequency of the low-pass filter, f_{CLP}.

f_{CHP}: _____

5. Create the spreadsheet table as shown below. Create the frequency column B by multiplying the low pass filter cutoff frequency, f_{CLP}, by the multipliers in column A. Create the frequency column D by multiplying the high-pass filter cutoff frequency, f_{CHP}, by the multipliers in column A.

	A	B	C	D	E	F
1	Multiplier	Frequency	V_{LP} V p-p	Frequency	V_{HP} V p-p	$V_{LP} + V_{HP}$
3	0.01					
4	0.1					
5	0.25					
6	0.5					
7	1					
8	2					
9	4					
10	10					
11	100					

6. Connect channel 2 of the oscilloscope to measure V_{LP}. Measure the magnitude of V_{LP} for each frequency in column B and record the result in column C.

7. Connect channel 2 of the oscilloscope to measure V_{HP}. Measure the magnitude of V_{HP} for each frequency in column D and record the result in column E.

Part B: The Sum

1. Connect the outputs of the filters together using 100K resistors as shown in the diagram on the right.

   Connect channel 2 of the oscilloscope to the combined output, "$V_{HP} + V_{LP}$".

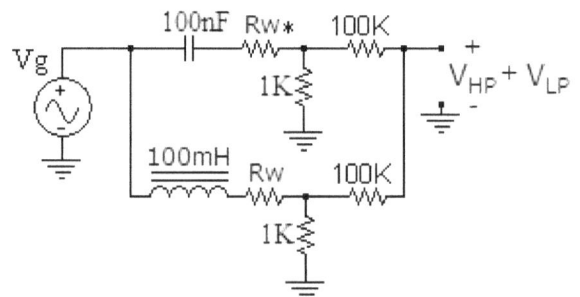

2. Use the frequencies in the spreadsheet table column B to measure the magnitude of VHP + VLP and record the results in column F. Do not measure the phase angles (they should be close to zero degrees).

   Note that the amplitudes should be approximately constant over the entire frequency range. The amount of variation depends on how close the cutoff frequencies of the two filters are to each other.

Analysis Note

It is useful to remember that the cutoff frequency, ω_c, for simple filters can be calculated from the "time constant" of the components as simply the reciprocal of the time constant. This makes sense unit-wise because frequency is the reciprocal of time.

For an RC filter circuit: $\tau = RC$ For an RL filter circuit: $\tau = \dfrac{L}{R}$.

Given that $\omega = 2\pi f$, the cutoff frequencies for both low-pass and high-pass RC filters can be expressed as:

$$\omega_c = \frac{1}{\tau} = \frac{1}{RC} \text{ and } f_c = \frac{1}{2\pi RC}.$$

The cutoff frequencies for both low- and high-pass RL filters can be expressed as:

$$\omega_c = \frac{1}{\tau} = \frac{R}{L} \text{ and } f_c = \frac{R}{2\pi L}.$$

Simulation results for each filter and the combined output of both filters are shown below. Note that the combined response is almost flat with a slight peaking near the cutoff frequencies.

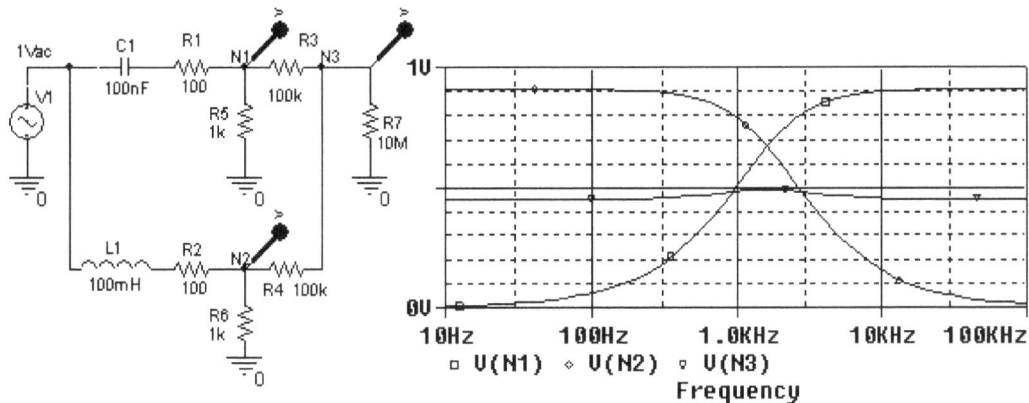

Analysis, Part A

1.   Use the spreadsheet to plot the amplitude versus frequency for the high-pass filter (column C) and the low-pass filter (column E). Use log scales.

2.   Estimate the cutoff frequencies and compare the result to theoretical calculations.

3.   Explain how the resistance of the inductor affects the filter's cutoff frequency and the filter's band-pass output amplitude.

Analysis, Part B

1.   Use the spreadsheet (column F) to plot the amplitude versus frequency response for the combined output of both filters of the crossover network. Use log scales.

2.   Explain why the magnitude of the combined response of both filters is about one half the magnitude of the pass band response of each individual filter. Hint: use superposition.

117

Experiment 37: Two-Pole Low-Pass and Band-Pass Filters

Introduction

Single pole RC and RL filters may be cascaded to obtain more attenuation beyond the cutoff frequency. For example, a 2-pole low-pass filter made by cascading two single-pole filters with the same cutoff frequency will ideally have an attenuation of 40dB per decade beyond the cutoff frequency, which is 20dB more per decade than that of a single-pole filter.

However, two such filters connected together will affect each other so that the attenuation past the cutoff frequency will be less than ideal. In this exercise, the impedance of the first filter section is much lower than the second. This minimizes the source resistance effect of the first filter on the second, and the loading effect of the second filter on the first.

The cutoff frequency of the cascaded filter will be different than the cutoff frequency of the individual filter sections. This is because each filter section has an attenuation of 3dB at its cutoff frequency so that two cascaded sections will have an attenuation of 6dB at that frequency.

Single pole RC and RL filters may also be cascaded to obtain a band-pass or band-stop response. In this exercise, a low impedance high-pass filter is connected to a high impedance low-pass filter to obtain a wide bandwidth band-pass response.

Objectives

The frequency response of a two-pole low-pass filter and of a two-section band-pass filter will be measured. Measured results will be compared to simulation. In addition, the cascaded filter responses will be compared to the response of the individual filter sections.

Procedure

Equipment and Parts
Function generator, oscilloscope
Resistors: 2.2K, 22K, ¼ watt, 5%.   Capacitors: 1nF, 47nF, 470nF, 5%.

Part A: 2-Pole Low Pass Filter

1. Connect the circuit on the right. Connect the oscilloscope channel 1 to Vg, the function generator. Connect channel 2 to measure Vo. Set the trigger to channel 1. Set the function generator to produce a 100Hz, 8V peak-to-peak sine wave.

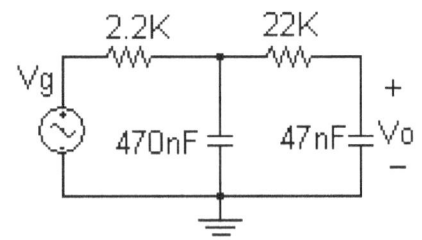

2. Tune the function generator to find the cutoff frequency, fc, where Vo is 0.707 of the maximum value. This occurs at the frequency where the magnitude of Vo = 5.67V ($Vo=8/\sqrt{2}$).

Create the spreadsheet table below to record the measured amplitude and phase angle of Vo for the frequencies specified. Enter "=A3*fc" into cell B3 using your cutoff frequency for fc to have the spreadsheet calculate the frequencies in column B.

	A	B	C	D	E	F	G	H
1	Mult.	Freq.	Measured Vo(mag)	Measured Vo(angle)	Calculated Vo(mag)	Calculated Vo(angle)	% Error Vo(mag)	% Error Vo(angle)
3	.1							
4	.5							
5	1							
6	2							
7	4							
8	10							
9	100							

Procedure Part B: Band-Pass Filter

1. Connect the circuit on the right. Connect the oscilloscope channel 1 to Vg, the function generator. Connect channel 2 to measure Vo. Trigger to channel 1.

    Set the function generator to produce a 1kHz, 8V p-p amplitude sine wave.

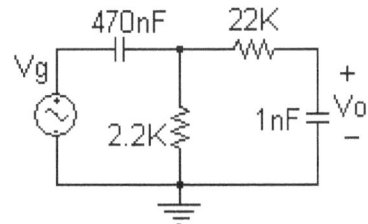

2. Create the spreadsheet table on the right to record the measured amplitude and phase angle of Vo for the frequencies specified. First fine-tune the function generator to find the filter's maximum output voltage, Vo(max). This should occur at about 1kHz.

    In row 8 of the spreadsheet, record the measured frequency, magnitude, and phase angle of Vo(max).

	A	B	C	D
1	Frequency		Measured Vo(mag)	Measured Vo(angle)
3	.01f1			
4	.10f1			
5	.50f1			
6	f1			
7	2f1			
8	Vo(max)			
9	.5f2			
10	f2			
11	2f2			
12	10f2			
13	100f2			

3. Find the low cutoff frequency f1 and record it into the spreadsheet cell B6. Measure and record the magnitude and phase angle of Vo at the frequency f1.

4. Find the high cutoff frequency f2. Record it into the spreadsheet cell B10. Measure and record the magnitude and phase angle of Vo at the frequency f2.

5. Use the low cutoff and high cutoff frequencies to calculate the other frequencies in the table. Measure and record the magnitude and phase angle of Vo at these calculated frequencies.

Analysis, Part A

1. Simulate the two-pole low-pass filter circuit. Refer to the simulation example at the end of this lab.

2. Enter the results of the simulation into columns E and F of your spreadsheet. Have the spreadsheet calculate the percent error in the magnitudes column G. Have the spreadsheet calculate the percent error in the phase angles column H.

3. Convert the magnitudes in column C to dBV by entering the equation "=20*LOG10(E3/(2*SQRT(2)))" into cell I3.

   Use the spreadsheet's pull down feature to calculate the dBV amplitudes for column I. Plot the magnitude response of your filter in dBV (column I versus column B).

4. Plot the phase response of your filter (column D versus column B). Refer to the spreadsheet example below.

Analysis, Part B

1. Use the spreadsheet to plot the magnitude response in dBV and phase response in degrees of the band-pass filter. Refer to the spreadsheet plots in analysis part A.

2. Simulate the band-pass filter circuit. Refer to the simulation example at the end of this lab. Plot the magnitude response and use the cursors to find the cutoff frequencies, f1 and f2, and the bandwidth of the filter.

3. Calculate percent difference between measured and simulated cutoff frequencies.

Spreadsheet Example

	A	B	C	D	E	F	G	H	I
1	Frequency		Measured Vo(mag)	Measured Vo(angle)	Calculated Vo(mag)	Calculated Vo(angle)	% Error Vo(mag)	% Error Vo(angle)	dBV(rms)
3	0.1	9.2			7.966	-7.180			8.993906
4	0.5	46			7.233	-34.571			8.155469
5	1	92			5.573	-62.880			5.890881
6	2	184			3.141	-99.700			0.910459
7	4	368			1.161	-133.200			-7.73426
8	10	920			0.217	-160.100			-22.3017
9	100	9200			0.00224	-178.000			-62.0259

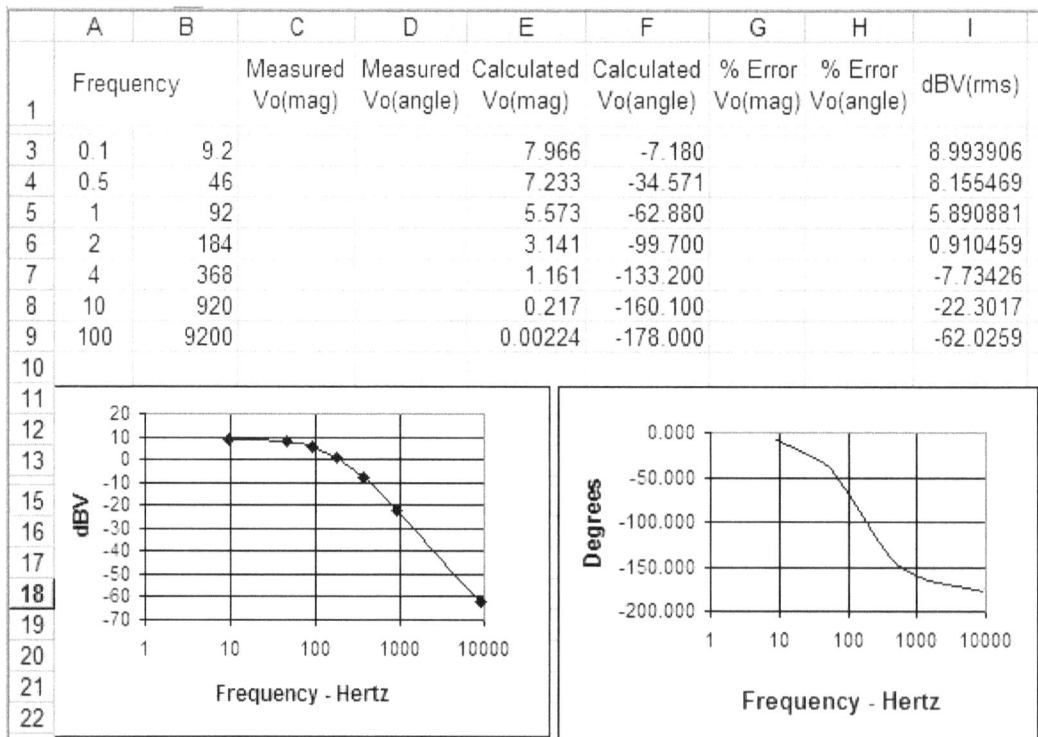

120

LTspice AC Sweep Simulation

The frequency response of three filters was simulated simultaneously by connecting them to the same source. The band-pass filter has a maximum response at about 1kHz and a bandwidth of about 10kHz. Also the 2-pole low-pass filter frequency response (node N2) is compared to the single-pole low-pass filter response (node N3).

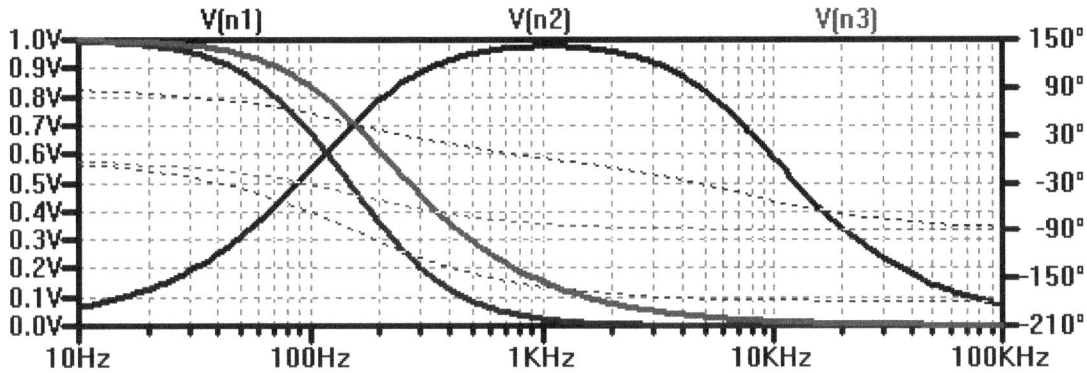

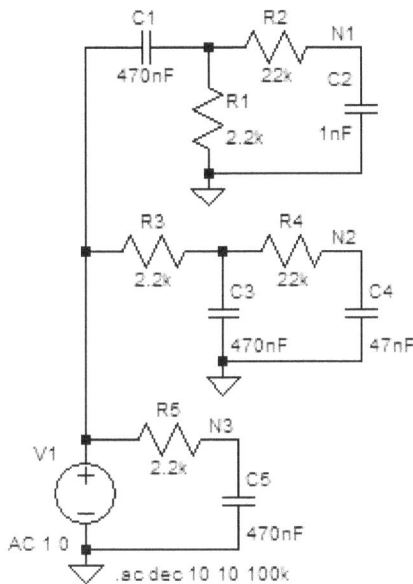

The magnitude axis was changed to linear by left clicking on it and selecting "Manual Limits". Select "Linear" and a range of 0 to 1V.

Left click on a voltage node at the top of the graph to obtain cursors for it. These are used to fid the magnitude and phase angle of the response at selected frequencies.

Two cursors may be displayed simultaneously by selecting "1st & 2nd" in the "Attach Cursor" dialog box.

Experiment 38: Series Resonant Passive Band-Pass Filter

Introduction

The impedance of a series RLC is a function of frequency. Inductive reactance is directly proportional to frequency, while the capacitive reactance is inversely proportional to frequency. Refer to the equations and circuit below.

$$X_L = 2\pi f L \qquad X_C = \frac{1}{2\pi f C} \qquad \mathbf{Z} = R + \mathbf{j}(X_L - X_C)$$

At the circuit's resonant frequency its net reactance is zero, $(X_L - X_C = 0)$ and its impedance, $\mathbf{Z}$, is equal to R.

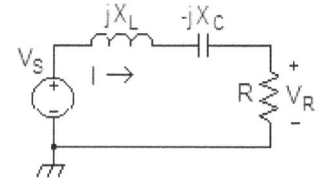

In this experiment, the circuit's resonant frequency and bandwidth will be measured. The results will be compared to theory.

Series Resonance

An example series RLC circuit is shown on the right. R_W is the resistance of the inductor. The circuit's transfer function is given below.

$$\mathbf{T}_s = \frac{\mathbf{Vo}}{\mathbf{V}_s} = \frac{R}{(R_w + R) + \mathbf{j}(2\pi f L - \frac{1}{2\pi f C})}.$$

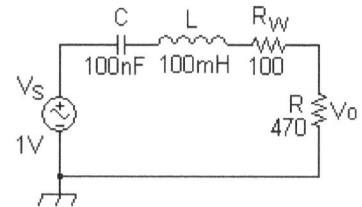

The circuit's resonant frequency, f_0, and transfer function at its resonant frequency, $\mathbf{T}_{s0}$, can be calculated by setting X_L equal to X_C.

$$f_0 = \frac{1}{2\pi\sqrt{LC}} = 1592\,Hz., \qquad \mathbf{T}_{s0} = \frac{R}{R_w + R} = 0.825.$$

The bandwidth, BW, of the resonant circuit is defined as the difference between the upper and lower cutoff frequencies, f_2 and f_1, of the circuit. These two frequencies are defined as the frequencies where the value of the transfer function is equal to $1/\sqrt{2}$ of its value at resonance. These frequencies are also called the -3dB frequencies because $1/\sqrt{2}$ is equal to -3dB.

As a band-pass filter, this circuit will pass frequencies between f1 and f2 with an amplitude greater than $1/\sqrt{2}$ of the maximum value which occurs at the frequency, f_0.

Another important parameter of a resonant circuit is its quality factor, Q. It is defined as the ratio of the energy stored in the circuit to the energy dissipated by the circuit per cycle.

It can be shown that the circuit's bandwidth, BW, is related to Q and f_0 by the equations below:

$$Q = \frac{X_L}{R + R_w} = \frac{2\pi f_0 L}{R + R_w}, \qquad BW = \frac{f_0}{Q} = \frac{R + R_w}{2\pi L} = \frac{570}{2\pi(0.1)} = 907 \ Hz.$$

Note that the bandwidth in units of radians per second would be $(R + R_W)/L$. This is just the ratio of the circuit's resistance to its inductance.

The cutoff frequencies of an RLC band-pass filter can always be calculated with the equation:

$$f_1, f_2 = \frac{1}{2\pi}\left[\pm\left(\frac{R}{2L}\right) + \frac{1}{2}\sqrt{\left(\frac{R}{L}\right)^2 + \frac{4}{LC}}\right] \text{(Hz)}$$

In this equation, R refers to the resistance of the entire circuit, and is equal to (R1 or R2) + Rw. If a filter's quality factor, Q, is greater than 10, the cutoff frequencies will be approximately bisected, or split down the center, by the center frequency. Mathematically, this can be expressed as:

$$f_1, f_2 = f_0 \pm \frac{BW}{2}$$

Of course, the equation immediately above is preferred because it is simpler, but Q must be evaluated to determine whether the simpler or more complex version should be used.

This circuit's transfer function has a maximum value of 0.825. Its bandwidth is the range of frequencies where the transfer function's value is greater than 0.583. This is from about 1100 Hz to 2000 Hz on the graph below.

A graph of the frequency response of the circuit on the previous page is shown on the right.

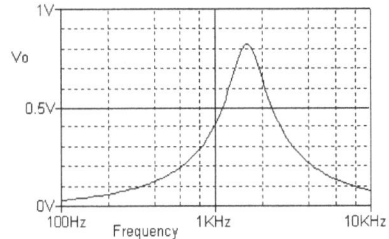

Note from the equations above that the circuit's bandwidth is directly proportional to the value of the circuit's resistance.

Procedure

Equipment and Parts
Function Generator, Oscilloscope with 10X probes, and Breadboard. Inductor: 100mH, Capacitor: 1nF, both 5%. Resistors: R1 = 2200Ω, R2 = 4700Ω, both ¼ watt, 5%.

1. Measure and record the resistance of the resistors, R1 and R2, and Inductor, Rw. If possible, measure the inductance of the inductor and capacitance of the capacitor

    R1$_{2200}$: _____   R2$_{4700}$: _____   C: _____

    L: _____   Rw: _____

2. Connect the circuit on the right. Connect C, L, and R with no wires in between.

    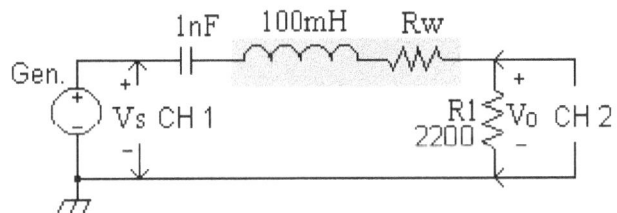

3. Set oscilloscope channel 1 to measure Vs.

    Set the function generator to produce a 16kHz, 5V peak-to-peak sine wave, with no offset on channel 1 of the oscilloscope. Set the oscilloscope to trigger on channel 1. Set channel 2 to measure Vo.

4. Connect the function generator to the circuit. Fine-tune the generator to the frequency where Vo is maximum. Check that Vs = $5\angle 0°$ V p-p for steps 4 through 8.

5. Record the frequency and the magnitude of Vo. Measure and record the phase angle, θ_0, of Vo with respect to Vs.

   f_0: _____   Vo(max): _____ θ_0: _____

6. Adjust the function generator to a frequency below f_0 where Vo's magnitude is 0.707 of its maximum value. Record the frequency f_1. Measure and record the phase angle θ_1.

   f_1: _____   Vo(-3dB): _____ θ_1: _____

7. Adjust the function generator to a frequency above f_0 where Vo's magnitude is 0.707 (-3dB) of its maximum value. Record the frequency f_2. Measure and record angle θ_2.

   f_2: _____   Vo(-3dB): _____ θ_2: _____

8. Replace the 2200Ω resistor with the 4700Ω resistor and repeat steps 3 through 7. Record your results below:

   f_0: _____   Vo(max): _____ θ_0: _____

   f_1: _____   Vo(-3dB): _____ θ_1: _____

   f_2: _____   Vo(-3dB): _____ θ_2: _____

Analysis

1. Organize your data into a spreadsheet table such as the one below:

	A	B	C	D	E	F	G	H	I	J
1	Measured R = 2200 ohms					Simulated R = 2200 ohms				
3		Freq. Hz	Vo p-p	Vo Deg.			Freq. Hz	Vo p-p	Vo Deg.	% Error
4	Fo						Fo			
5	F1						F1			
6	F2						F2			
7	BW						BW			
9	Measured R = 4700 ohms					Simulated R = 4700 ohms				
11		Freq. Hz	Vo p-p	Vo Deg.			Freq. Hz	Vo p-p	Vo Deg.	% Error
12	Fo						Fo			
13	F1						F1			
14	F2						F2			
15	BW						BW			

Use the measured values of your parts to do the calculations and simulations below.

2. Transfer your measured results into the table. Calculate both circuits' theoretical resonant frequencies and enter them into the table. Use these values to calculate the both circuits' quality factors. Based on the quality factors, determine the best equation for finding the theoretical cutoff frequencies. Calculate the theoretical values of the cutoff frequencies and enter them into the table.

3. Enter equations into the appropriate cells to calculate BW and percent error.

4. Simulate the circuit to check the results of your calculations. Refer to the simulation examples on the following pages.

5. Optional: Design a series resonant band-pass filter with a resonant frequency of 50kHz and bandwidth of 10kHz, using a 22mH inductor with a winding resistance of 20Ω. Simulate the design and write a short report on the results.

LTspice Example: Series Resonant Band-Pass Filter

This example uses a linear frequency sweep from 10kHz to 25kHz. V1 signal source magnitude is set to 5, and angle is set to 0. The output voltage magnitude and phase is plotted on the linear axis.

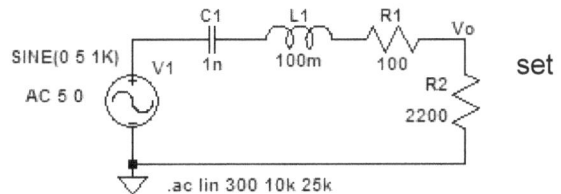

Set V1 to: SINE, *DC Offset* = 0, *Amplitude* = 5, *Freq.* = 1k, *AC Amplitude* = 5, *AC Phase* = 0
Click on Simulate in the main menu and select *Edit Simulation Cmd.* Select "AC Analysis". set:
Type of Sweep = Linear, *Number of Points* = 300, *Start Freq.* = 10k, and *Stop Freq.* = 25k.

The resulting Bode plot below shows a resonant frequency of about 16kHz and a bandwidth of about 3200Hz. The phase angle is 0 degrees at the resonant frequency, 45 degrees at the lower cutoff frequency, and -45 degrees at the upper cutoff frequency.

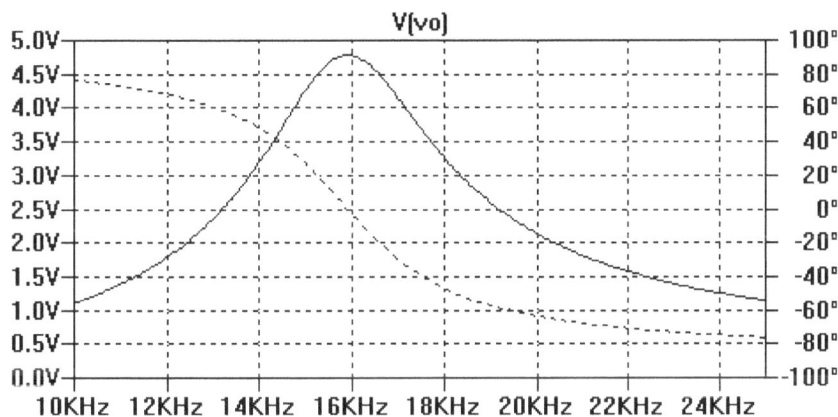

Vo and θ as a Function of Frequency

Experiment 39: Parallel Resonant Band-Pass Filter

Introduction

Determining the effect of inductor winding resistance is somewhat more difficult in a parallel resonant circuit. However, the effect on the circuit's quality factor, Q, and bandwidth, BW, can be approximated by converting the series resistance of the inductor winding to an equivalent parallel resistance.

The measured response of the parallel RLC circuit will be compared to analysis and simulation. Two important parameters, resonant frequency, f_0, and bandwidth, BW, will be measured.

Parallel Resonance

A simple parallel RLC circuit is shown on the right. As with the series RLC circuit, its resonant frequency, f_0, can be calculated by setting X_L equal to X_C:

$$f_0 = \frac{1}{2\pi\sqrt{LC}} = 1592\,Hz.$$

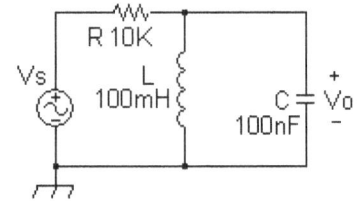

The circuit's quality factor, Q, as defined by the ratio of the energy stored to the energy dissipated per cycle, is R/X_L, where X_L is the reactance of the inductor at the circuit's resonant frequency.

A series-parallel RLC circuit is shown on the right. R_W is the winding resistance of the inductor. Consider the loop consisting of R_W, L, and C as a series resonant circuit. Calculate its resonant frequency and use the result to calculate the impedance of the series combination of R_W and L: $\mathbf{Z} = R_W + \mathbf{j}\,X_L$. Transform $\mathbf{Z}$ into a parallel equivalent circuit:

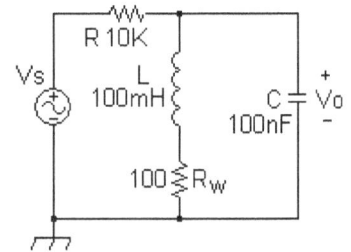

$$\mathbf{Y} = \frac{1}{\mathbf{Z}} = \frac{1(R_W - \mathbf{j}X_L)}{(R_W + \mathbf{j}X_L)(R_W - \mathbf{j}X_L)} = \frac{(R_W - \mathbf{j}X_L)}{R_W^2 + X_L^2} = \frac{R_W}{R_W^2 + X_L^2} - \mathbf{j}\frac{X_L}{R_W^2 + X_L^2}$$

Next convert the components of the admittance into components of parallel impedance:

$$R_P = \frac{R_W^2 + X_L^2}{R_W} \qquad X_P = \frac{R_W^2 + X_L^2}{X_L} \qquad L_P = \frac{X_P}{2\pi f_0}$$

The series-parallel circuit may be converted to an entirely parallel circuit by also converting the voltage source in series with R to a current source in parallel with R, as shown on the right.

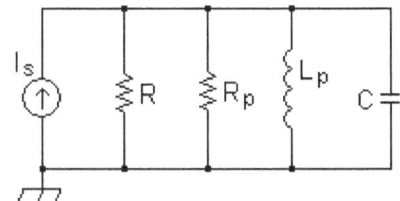

The circuit's Q and bandwidth can now be calculated from the parallel equivalent circuit:

$$Q = \frac{R \| R_P}{X_P} = \frac{R\,R_P}{2\pi f_0\, L_P(R + R_P)} \qquad BW = \frac{f_0}{Q} \qquad \text{(note that usually } L_P \approx L)$$

126

Procedure

> **Equipment and Parts**
>
> Function Generator, Oscilloscope with 10X probes, and Breadboard.
> Inductor: 100mH, 5%, Capacitor: 0.1µF, 5%.
> Resistor: R = 10K, ¼ watt, 5%.

Resonant Frequency and Bandwidth

1. Measure and record the resistance of the resistor, R, and inductor, Rw. If possible, measure the inductance of the inductor and capacitance of the capacitor.

    R: _____   C: _____

    L: _____   Rw: _____

2   Connect C, L, and R together with no wires in between. Connect the generator and oscilloscope leads directly to the components.

3. Set channel 1 of the oscilloscope to measure Vs. Set channel 2 to measure Vo. Set the oscilloscope to trigger on channel 1.

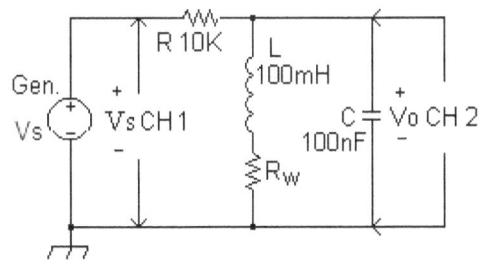

4. Set the function generator to produce a 5V peak-to-peak, 1600Hz sine wave, with no offset (this is Vs). Adjust the generator to the frequency where Vo is maximum. This will occur at the circuit's resonant frequency, f_0.

5. Measure and record the frequency, f_0, and the peak-to-peak value of the voltage, Vo. Also measure and record the phase angle, θ_0.

    f_0: _____   Vo(max): _____ θ_0: _____

6. Adjust the function generator to a frequency below f_0 where Vo is 0.707 of its maximum value. Measure the phase angle of Vo with respect to Vs. Record the frequency as f_1, and phase angle as θ_1.

    f_1: _____   Vo(-3dB): _____ θ_1: _____

7. Adjust the function generator to a frequency above f_0 where Vo is 0.707 of its maximum value. Measure the phase angle of Vo with respect to Vs. Record the frequency as f_2, and the phase angle as θ_2.

    f_2: _____   Vo(-3dB): _____ θ_2: _____

127

Analysis

1.   Organize your data into a spreadsheet table such as the one below:

	A	B	C	D	E	F	G	H	I	J
1	Measured					Simulated				
3		Freq. Hz	Vo p-p	Vo Deg.			Freq. Hz	Vo p-p	Vo Deg.	% Error
4	Fo					Fo				
5	F1					F1				
6	F2					F2				
7	BW					BW				

Use the measured values of your parts to do the calculations and simulations.

2.   Transfer your measured results into the table. Calculate the theoretical values of the resonant frequencies and cutoff frequencies. Enter the results into the table.

3.   Enter equations into the appropriate cells to calculate BW and percent error.

4.   Simulate the circuit to check the results of your calculations. Refer to the simulation example below.

LTspice Example: Parallel Resonant Circuit

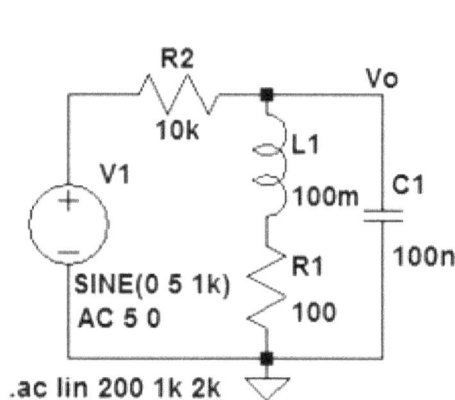

Circuit Diagram

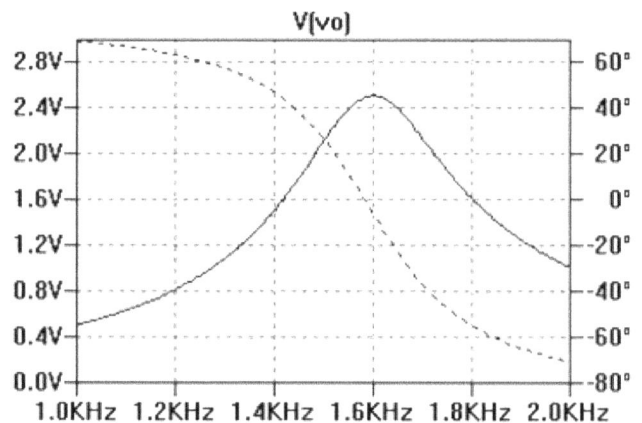

Magnitude and Phase Response

Experiment 40: Audio Output Transformer

Introduction

This lab exercise demonstrates the voltage, current, and impedance transformation properties of a transformer. An ideal transformer's transformation ratio is related to the turns, n, the voltage, V, and the current, I, in its primary and secondary coils by the following equations.

$$a = \frac{n_p}{n_s} = \frac{V_p}{V_s} = \frac{I_s}{I_p}$$

Recall that in order to achieve maximum power transfer to a resistive load, or to get the most power from a source to a load, the load must have the same resistance as the voltage source's series resistance. A transformer can be useful in making a load "appear" to be larger or smaller. The impedance seen by the terminals of a transformer's primary coil, Z_p, will be related to the actual resistance on the secondary coil, Z_s, and the transformation ratio by the following equations.

$$Z_p = a^2 Z_s \qquad\qquad a = \sqrt{\frac{Z_p}{Z_s}}$$

Z_p is called reflected impedance. In this experiment, an AC voltage source with a source impedance of 1000Ω is first connected directly to an 8.2Ω resistor, and the power delivered to it is measured. Then the same load resistance is connected to the source through an impedance matching transformer, and the power delivered to the 8.2Ω resistor is again measured. In addition, the primary and secondary voltages and currents will be determined and compared to the voltage transformation properties of the transformer.

An audio output transformer is used which is designed to match a high impedance audio amplifier output to a low impedance loudspeaker. An 8Ω loudspeaker will replace the 8.2Ω resistor, and the measurements will be repeated. This will also provide an audible demonstration of the effectiveness a transformer's ability to transfer more power to a load. The speaker will get louder as it converts more electric energy into sound energy.

In the second part of this exercise the transformer primary will be connected as an "auto-transformer", and its voltage and impedance matching properties will be explored. The autotransformer will be center-tapped, so it will have twice as many turns on its primary coil as its secondary coil, giving it a transformation ratio of 2.

Procedure

Equipment and Parts
Function Generator, Oscilloscope, DMM, and Breadboard. Audio Transformer, 1000Ω, center tapped, to 8Ω, 200mW. Loudspeaker, 8Ω, 200mW minimum, (2 to 4 inch). Resistors: R1: 1kΩ, R2: 8.2Ω, R3: 220Ω, R4: 47Ω, R5: 470Ω, ¼ watt, 5%.

Part A: Impedance Transformation

1. Measure the values of the resistors for use in your analysis.

   R1$_{1k}$: _____   R2$_{8.2}$: _____   R3$_{220}$: _____

   R4$_{47}$: _____   R5$_{470}$: _____

2. Connect the circuit on the right. Connect channel 1 of the oscilloscope to Vg, the function generator. Connect channel 2 of the oscilloscope to measure Vo.

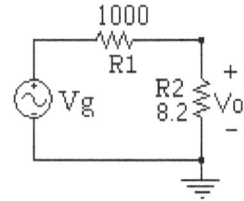

   Set the oscilloscope to trigger on channel 1. Set the function generator to produce a 10V peak-to-peak, 1000Hz sine wave with no offset.

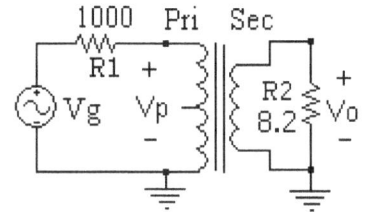

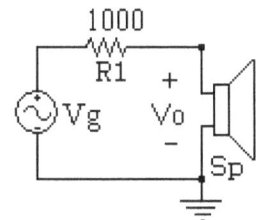

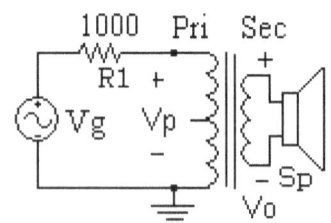

3. Measure and record the magnitude and phase of the voltage Vo.

   Vo(mag): _____ V p-p   Vo(angle): _____ Degrees

4. Connect the circuit on the right. Connect channel 1 of the oscilloscope to Vg, the function generator. Connect channel 2 of the oscilloscope to measure Vo. Set the oscilloscope to trigger on channel 1. Set the function generator to produce a 10V peak-to-peak, 1000Hz sine wave with no offset.

5. Measure and record the magnitude and phase of the voltage Vo.

   Vo(mag) _____ volts p-p.   Vo(angle) _____ degrees.

6. Measure and record the magnitude and phase of the voltage Vp.

   Vp(mag) _____ volts p-p.   Vp(angle) _____ degrees.

7. Measure and record the resistance of the loudspeaker.   Sp _____ohms.

8. Connect the circuit on the right. Connect channel 1 of the oscilloscope to Vg, the function generator. Connect channel 2 of the oscilloscope to measure Vo. Set the oscilloscope to trigger on channel 1. Set the function generator to produce a 10V peak-to-peak, 1000Hz sine wave with no offset.

9. Measure and record the magnitude and phase of Vo.

   Vo(mag) _____ volts p-p.   Vo(angle) _____ degrees.

10. Connect the circuit on the right. Connect channel 1 of the oscilloscope to Vg, the function generator. Connect channel 2 of the oscilloscope to measure Vo. Set the oscilloscope to trigger on channel 1. Set the function generator to produce a 10V peak-to-peak, 1000Hz sine wave with no offset.

11. Measure and record the magnitude and phase of Vo.

Vo(mag) _____ volts p-p.   Vo(angle) _____ degrees.

12. Measure and record the magnitude and phase of the voltage Vp.

Vp(mag) _____ volts p-p.   Vp(angle): _____ degrees.

Part B: Autotransformer

1. Connect the circuit on the right with R_L = 47Ω. Connect channel 1 of the oscilloscope to Vg, the function generator. Connect channel 2 of the oscilloscope to measure Vo.

   Set the oscilloscope to trigger on channel 1. Set the function generator to produce a 10V peak-to-peak, 1000Hz sine wave with no offset.

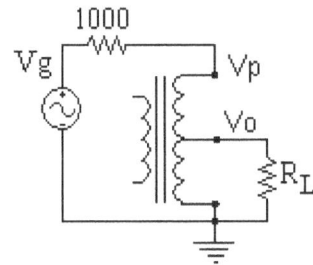

2. Measure and record the magnitude and phase of the voltage Vo.

Vo(mag) _____ volts p-p.   Vo(angle) _____ degrees.

3. Measure and record the magnitude and phase of the voltage Vp.

Vp(mag) _____ volts p-p.   Vp(angle) _____ degrees.

4. Repeat step 1 with R_L = 220Ω.

5. Measure and record the magnitude and phase of the voltage Vo.

Vo(mag) _____ volts p-p.   Vo(angle) _____ degrees.

6. Measure and record the magnitude and phase of the voltage Vp.

Vp(mag) _____ volts p-p.   Vp(angle) _____ degrees.

7. Repeat step 1 with R_L = 470Ω.

8. Measure and record the magnitude and phase of the voltage Vo.

Vo(mag) _____ volts p-p.   Vo(angle) _____ degrees.

9. Measure and record the magnitude and phase of the voltage Vp.

Vp(mag) _____ volts p-p.   Vp(angle) _____ degrees.

Analysis, Part A

1. Use the measurements in step 3 to calculate power delivered to the 8.2Ω load resistor. Compare measured power to the theoretically expected power. Calculate the ratio of the power delivered to the load resistor to the power delivered by the source.

2. Use the measurements in step 5 to calculate power delivered to the 8.2Ω load resistor. Compare measured to theoretically expected power. Calculate ratio of load resistor power to power delivered by the source (also use results of step 6).

131

3. Use the measurements in steps 5 and 6 to calculate the voltage transformation ratio of the transformer, and compare these measured values to the theoretically expected value. The theoretical value of the transformation ratio can be calculated from the impedance specifications of the transformer. More specifically, the package specifies that when there is an 8 ohm load on the secondary coil, 1000 ohms will be reflected on the primary coil.

4. Explain the results for the phase angles between the source, Vg, and the voltages Vp and Vo in steps 5 and 6.

Analysis: Part B

1. Use the measurements in part B, steps 2 and 3, to calculate the power delivered to 47Ω load resistor. Compare the measured power to the theoretically expected power. Calculate the ratio of the power delivered to the load resistor to the power delivered by the source.

2. Use the measurements in part B, steps 5 and 6, to calculate the power delivered to 220Ω load resistor. Compare the measured power to the theoretically expected power. Calculate the ratio of the power delivered to the load resistor to the power delivered by the source.

3. Use the measurements in part B, steps 8 and 9, to calculate the power delivered to 470Ω load resistor. Compare the measured power to the theoretically expected power. Calculate the ratio of the power delivered to the load resistor to the power delivered by the source.

4. Compare and comment on the results of the ratio of power delivered to the load resistor to the power delivered by the source for the 47Ω, 220Ω, and 470Ω loads resistors.

LTspice Transformer Notes

In LTspice, a transformer can be simulated with two inductors that are magnetically coupled. To magnetically couple two inductors, a SPICE directive must be added by clicking on the ".op" button in the upper right corner. To couple inductors L1 and L2 with a magnetic coupling coefficient of one, enter "K1 L1 L2 1" and place it anywhere in the circuit.

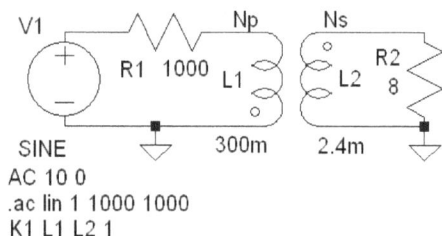

```
--- AC Analysis ---

frequency: 1000 Hz
V(np):  mag:      4.83287 phase:  14.8561°
V(ns):  mag:     0.432265 phase: -165.144°
V(n001) mag:          10 phase:       0°
I(R2):  mag:    0.0540331 phase: -165.144°
I(R1):  mag:   0.00547085 phase:  166.909°
I(V1):  mag:   0.00547085 phase:  166.909°
```

Experiment 41: Three Phase Power / Wye Connection

Introduction

This lab experiment requires a low-voltage 3-phase voltage source that can supply a 12V peak-to-peak sine wave at a current of at least 25mA rms per phase. One can be built on a breadboard as described in Appendix 2. A compact plug-in version of this 3-phase source is described in this lab experiment and also in Appendix 2.

The neutral connection of this source shares the same ground as the lab power supply, function generator, and oscilloscope. Therefore care must be taken to not ground a phase output, such as would occur if the ground lead of the oscilloscope were connected to one of the phase outputs. Study the diagram of a typical 120 VAC single-phase power system below. The grounds of the oscilloscope and function generator are connected together through the ground of the AC receptacles.

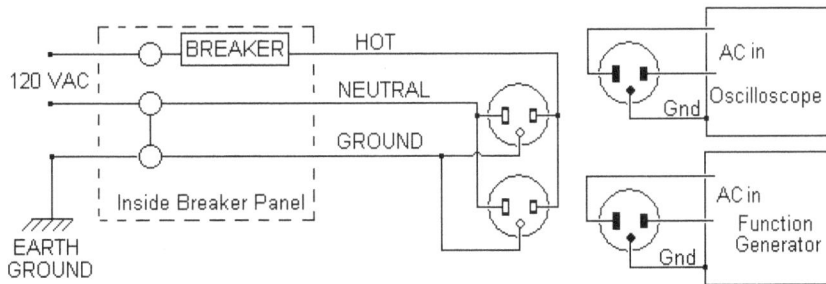

Note that the wide connector on the receptacle and the ground connector are connected together and to earth ground at the breaker panel. So the function generator and oscilloscope grounds are connected together through the AC power outlets.

This exercise involves the measurement of the magnitude and phase angles of voltages and currents in a 4-wire and 3-wire, 3-phase, wye connected power system. The effect of balanced and unbalanced loads with and without the neutral wire is measured.

Procedure

Equipment and Parts
Function Generator, Power Supply, 3-Phase Source, Oscilloscope, and Breadboard. Resistors: Four 100Ω, Four 470Ω, ¼ watt, 5%.

Part A: Balanced Load

1.   Measure the values of your resistors (refer to the diagram on the next page) and record:

R1		RA	
R2		RB	
R3		RC	
RG		RX	

If you already have a 3-phase voltage source, set it to produce a 60Hz, 12V peak-to-peak, sine wave per phase, as measured from each line output to neutral. The reference phase will be P0 in this experiment. The phase angle of P1 will be 120^0 with reference to P0 and the phase angle of P2 will be -120^0 with reference to P0.

Check that the voltages and phase angles of the 3-phase source are correct.

2.   If you don't already have a 3-phase voltage source, you can build the circuit described in appendix 2, or use a pc board version of the circuit, the *Phase Tripler*, or the *Phase Tripler II*, which are also described in appendix 2.

Connect the circuit as shown below. The circuit connected on a breadboard with the Phase Tripler II is shown on the right.

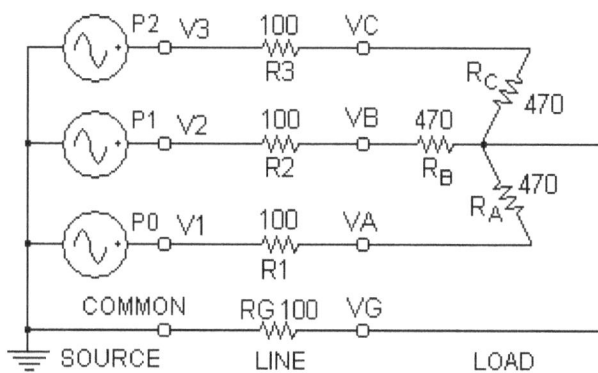

If you are using the *Phase Tripler II* board, handle it by the edges and carefully insert it into the breadboard. Connect the power supply ground and function generator ground to pin 2, function generator signal to pin 1, +12VDC to pin 4, and -12VDC to pin 6.

Connect the output pins to the circuit. Pin 9 is P0, pin 10 is P1, pin 11 is P2, and P12 is neutral (ground). Also connect the oscilloscope grounds to pin 12.

If you are using the *Phase Tripler*, the same inputs are required except banana binding posts are used for the DC inputs and phase outputs. The function generator is connected to the BNC.

3.   Connect channel 1 of the oscilloscope to P0 and set the trigger to channel 1. Set the function generator amplitude to produce a 12V peak-to-peak sine wave with no offset on channel 1.

4.   Connect channel 2 of the oscilloscope to P1. Check that the amplitude is exactly 12V peak-to-peak and that the phase is leading channel 1 by exactly 120^0.

5.   Connect channel 2 of the oscilloscope to P2. Check that the amplitude is exactly 12V peak-to-peak and that the phase is lagging channel 1 by exactly 120^0.

[Refer to the calibration procedure in Appendix 2 if steps 3 and 4 don't check.]

If you are using a different 3-phase source, record the magnitude and phase angle of the voltage supplying phases A, B, and C of the load circuit.

6. Measure the magnitude and phase angle of the voltages VG, VA, VB, and VC with channel 2. Record results below.

Step 6	VG	VA	VB	VC
Magnitude p-p				
Phase, Deg.				

7. Open the neutral line by removing the 100Ω resistor between the phase tripler ground and node VG.

8. Measure the magnitudes and phase angles of the voltages VG, VA, VB, and VC with the oscilloscope channel 2 and record results in the table below.

Step 8	VG	VA	VB	VC
Magnitude p-p				
Phase, Deg.				

Part B: Unbalanced Load

1. Connect a 470Ω resistor (RX) in parallel with RC so that the RC branch becomes 235Ω, as shown on the right. Reconnect the 100Ω resistor between node VG and COMMON.

2. Measure the magnitudes and phase angles of the voltages VG, VA, VB, and VC with oscilloscope channel 2 and record results in the table below.

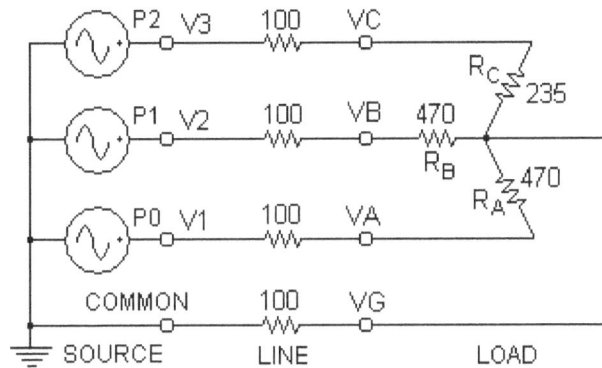

Step 2	VG	VA	VB	VC
Magnitude p-p				
Phase, Deg.				

3. Open the neutral line by removing the 100Ω resistor between the phase tripler ground and node VG.

4. Measure the magnitude and phase angle of the voltages at nodes nG, nA, nB, and nC with the oscilloscope channel 2 and record results in the table below.

Step 4	VG	VA	VB	VC
Magnitude p-p				
Phase, Deg.				

Analysis, Part A

1. Use your measured resistor values and the node voltage method to calculate the voltage VG for the circuit used in procedure part A. Use the result to calculate the current supplied by the source for each phase (both magnitude and phase angle).

2. Calculate the magnitudes and phases angles of the voltages VA, VB, and VC. Calculate the neutral wire current.

3. Calculate the percent error between the measured and calculated values of the voltage magnitudes VA, VB, and VC. Calculate the absolute error between the phase angles.

Analysis, Part B

1. Use LTspice to simulate the unbalanced circuit with and without the neutral connection. Refer to the example below.

2. Calculate the percent error between the measured and simulated values of the voltage magnitudes VA, VB, and VC. Calculate the absolute error between the phase angles.

3. Use Kirchhoff's current law to show that the neutral wire current is the result of the unbalance of the phase currents.

LTspice 3-Phase Analysis

Three phase circuits can be simulated using three voltage sources. Right click on each source and set their AC Amplitude and phase angle.

Use AC analysis with the start and stop frequencies set to 60Hz and the number of points set to 1.

Use your measured resistor values for your simulation.

The results of simulating the circuit on the right are given below. Verify for yourself that the current supplied by each source is 20mA p-p.

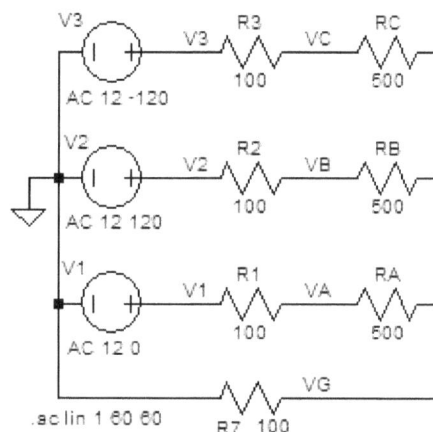

```
--- AC Analysis ---
frequency:    60          Hz
V(vc):        mag:    10 phase:        -120°        voltage
V(v3):        mag:    12 phase:        -120°        voltage
V(vb):        mag:    10 phase:         120°        voltage
V(v2):        mag:    12 phase:         120°        voltage
V(va):        mag:    10 phase:  6.36111e-016°      voltage
V(v1):        mag:    12 phase:           0°        voltage
```

Experiment 42: Three Phase Power / Delta Connection

Introduction

This lab exercise requires a 3-phase voltage source. Refer to the introduction and procedure steps 1 and 2 of experiment 41 and appendix 2 in this lab book.

This exercise involves the measurement of the magnitudes and phase angles of voltages and currents in a 3-wire, 3-phase, delta connected power system. The effect of balanced and unbalanced loads is measured.

Procedure

Equipment and Parts
Function Generator, Power Supply, 3-Phase Source, Oscilloscope, and Breadboard. Resistors: Four 390Ω, ¼ watt, 5%. Inductors: Three 100mH, 5% (iron core, 50mA minimum).

Part A: Balanced Load with Inductive Line

1. Measure the values of your resistors, inductors (if possible), and inductor resistances, (refer to the diagram on the next page) and record below:

Rw1		L1		RA	
Rw2		L2		RB	
Rw3		L3		RC	
				RX	

1. Connect the circuit and your 3-phase source as shown on the right.

    Physically separate the inductors from each other as much as possible on the breadboard to reduce unwanted magnetic coupling between inductors.

    $P0 = 12\angle 0^{0}$ V_{p-p}, $P1 = 12\angle 120^{0}$ V_{p-p}, $P2 = 12\angle -120^{0}$ V_{p-p}.

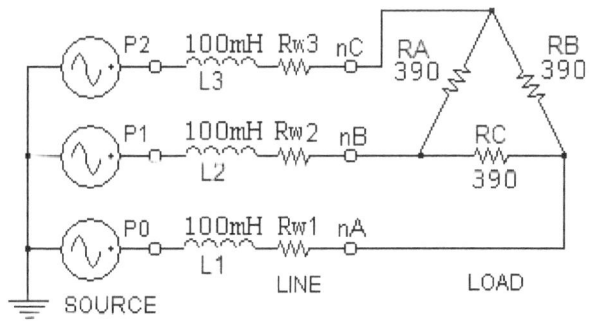

2. Connect channel 1 of the oscilloscope to P0 and set trigger to channel 1. Be sure that all instrument grounds are connected to the source ground.

3.	Measure the magnitude and phase angle of the voltages at nodes nA, nB, and nC with the oscilloscope channel 2 and record results in the table below.

Step 3	nA	nB	nC
Magnitude p-p			
Phase, Deg.			

Part B: Unbalanced Load with Inductive Line

1.	Connect a 390Ω resistor(RX) in parallel with RC so that the RC branch becomes 195Ω.

2.	Measure the magnitude and phase angle of the voltages at nodes nA, nB, and nC with the oscilloscope channel 2 and record results in the table below.

Step 2	nA	nB	nC
Magnitude p-p			
Phase, Deg.			

3.	Optional: Repeat steps 1 and 2 above with a 4.7µF capacitor in parallel with RC instead of the 390Ω resistor, (RX).

Analysis

1.	Use your measured resistor values and the node voltage method to calculate the voltages at nodes nA, nB, and nC for the balanced circuit. Save the equation on your calculator for the next step. Refer to the TI-89 example at the end of this experiment.

2.	Use the node voltage method to calculate the voltages at nodes nA, nB, and nC for the unbalanced load circuit (edit your equation from step 1).

3.	Calculate the percent error between the measured and calculated magnitudes of the voltages at nodes nA, nB, and nC for the balanced and unbalanced loads. Calculate the absolute error between the phase angles.

4.	For both the balanced and unbalanced circuits, calculate the voltage across each load resistor using the measured results for the voltages at nA, nB, and nC.
	$[(V_{nA} - V_{nB}), (V_{nB} - V_{nC}), \text{ and } (V_{nC} - V_{nA})]$

5.	Optional: Present an analysis of the results for procedure Part B, step 3.

LTspice Example

Simulate the balanced and unbalanced circuit using your measured part values.

The results for the circuit on the right are presented below.

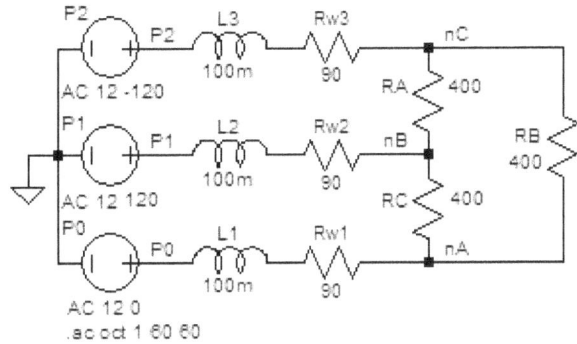

```
                    --- AC Analysis ---

    frequency:      60              Hz
    V(nc):          mag:    7.06421  phase:   -129.581°       voltage
    V(nb):          mag:    7.06421  phase:    110.419°       voltage
    V(na):          mag:    7.06421  phase:   -9.58128°       voltage
    V(p2):          mag:         12  phase:       -120°       voltage
    V(p1):          mag:         12  phase:        120°       voltage
    V(p0):          mag:         12  phase:          0°       voltage
```

TI-89 Example

The following example uses 330-ohm load resistors instead of 390-ohm resistors.

Let $V(nA) = x$, $V(nB) = y$, $V(nC) = z$.

Equations:

$$\frac{x-12\angle 0}{100+37.7i}+\frac{x-y}{330}+\frac{x-z}{330}=0 \text{ and } \frac{y-12\angle 120}{100+37.7i}+\frac{y-x}{330}+\frac{y-z}{330}=0 \text{ and } \frac{z-12\angle 240}{100+37.7i}+\frac{z-x}{330}+\frac{z-y}{330}=0$$

TI-89 Input for Balanced Case:

csolve((x-12)/(100+37.7i)+(x-y)/330+(x-z)/330=0 and (y-(12∠120))/(100+37.7i)+(y-x)/330+(y-z)/330=0 and (z-(12∠240))/(100+37.7i)+(z-x)/330+(z-y)/330=0,{x,y,z})

x = (6.187∠-10.8)   y = (6.187∠109.8)   z = (6.187∠-130.2)

TI-89 Input for Unbalanced Case:

csolve((x-12)/(100+37.7i)+(x-y)/330+(x-z)/220=0 and (y-(12∠120))/(100+37.7i)+(y-x)/330+(y-z)/330=0 and (z-(12∠240))/(100+37.7i)+(z-x)/220+(z-y)/330=0,{x,y,z})

x = (5.611∠-15.12)   y = (6.187∠109.82)   z = (5.477∠-127.29)

139

Experiment 43: Fourier Series and Circuit analysis

Introduction

According to Fourier analysis, a square wave may be considered to be a superposition of an infinite number of odd harmonic frequencies whose amplitudes decrease inversely with frequency. The fundamental is the lowest frequency of the square wave. The Fourier series of a square wave voltage, whose peak-to-peak amplitude is V, and average value is V_{AVE}, may be expressed as:

$$v(t) = V_{AVE} + \frac{2V}{\pi} \sum_{n}^{\infty} \frac{1}{n} \sin(2\pi n f_0 t), \text{ where n is an odd}$$

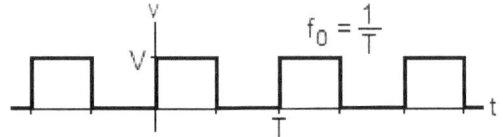

integer and the fundamental frequency is f_0.

If a square wave is applied to a filter circuit, the wave's frequency components will be affected by the filter's frequency response. The filter's output frequency spectrum will be different than the input frequency spectrum.

Although the Fourier series of any time domain waveform consists of an infinite number of harmonic frequencies, an adequate approximation may be obtained using only the first several terms of the series. The average value plus the first 5 terms for the square wave are:

$$v(t) = V_{AVE} + \frac{2V}{\pi} \sin(2\pi f_0 t) + \frac{2V}{3\pi} \sin(6\pi f_0 t) + \frac{2V}{5\pi} \sin(10\pi f_0 t) + \frac{2V}{7\pi} \sin(14\pi f_0 t) + \frac{2V}{9\pi} \sin(18\pi f_0 t)$$

Fourier analysis may be used to calculate the response of a filter to each frequency component of the input waveform using phasor analysis. First the magnitude and phase angle of each frequency component is calculated. Next the transfer function of the filter is calculated for each frequency component. Then the output of the filter at each frequency is calculated as the phasor product of the input and the transfer function.

An example is presented here for the filter circuit on the right. The input, Vg, is a 2V peak-to-peak, 100Hz square wave with a 1V offset. The first 4 terms of the Fourier series for Vg are:

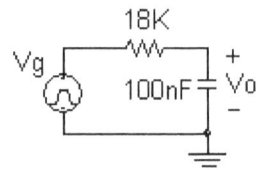

$$v_g(t) = 1 + 1.273 \sin(2\pi 100t) + .424 \sin(2\pi 300t) + .255 \sin(2\pi 500t)$$

The transfer function of the filter is: $T_f = \dfrac{Vo}{Vg} = \dfrac{1}{1 + j\dfrac{f}{88.3}}$.

Its magnitude is: $|T_f| = \dfrac{1}{\sqrt{1 + \left(\dfrac{f}{88.3}\right)^2}}$. Its phase angle is: $\theta = -\arctan\left(\dfrac{f}{88.3}\right)$.

The transfer function of the filter in phasor form for each frequency, DC, 100Hz, 300Hz, and 500Hz is:

$V_{DC} = 1,$ $\qquad V_{100} = .662\angle{-48.56},$ $\quad V_{300} = .282\angle{-73.6},$ $\quad V_{500} = .174\angle{-80.0}.$

140

The output of the filter can now be calculated:

$V_{DC}=1(1)$, $V_{100}=(1.273\angle0^0)(.662\angle-48.56^0)$, $V_{300}=(.424\angle0^0)(.282\angle-73.6^0)$, $V_{500}=(.255\angle0^0)(.174\angle-80.0^0)$

$v_o(t) = 1 + .843\sin(2\pi100t - 48.56^0) + .120\sin(2\pi300t - 73.6^0) + .044\sin(2\pi500t - 80.0^0)$

Compare the above results to the simulation below. The plot on the left below shows the output waveform of the filter in the time domain. The plot on the right below shows the spectrum of the square wave input, V1, and the spectrum of the output of the filter.

The magnitudes of the harmonics are given in peak units

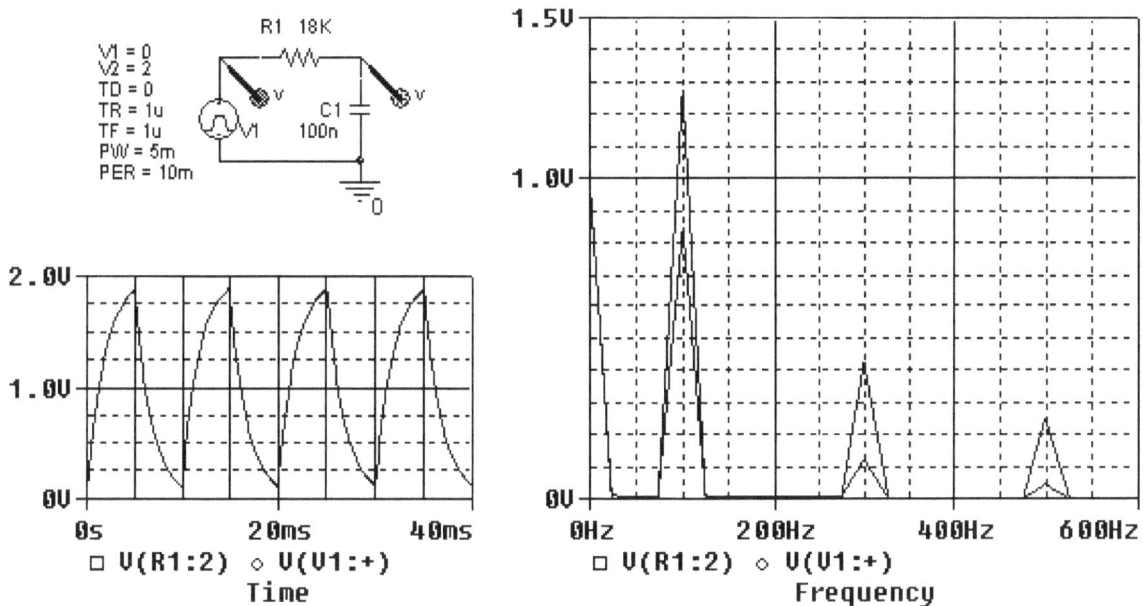

Procedure

Equipment and Parts

Function Generator, Oscilloscope, and Breadboard.
Resistor: 15K, ¼ watt, 5%. Capacitor: 100nF, 5%.

1. Connect the circuit on the right. Connect the oscilloscope channel 1 to node N1 and channel 2 to node N2.

2. Vg is the function generator set to produce a 120Hz, 3V peak-to-peak, square wave with a 1.5 volt offset (goes between 0 and 3 volts).

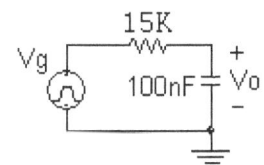

3. Observe the time domain waveform on channel 2. Capture or sketch about 2 cycles of the output waveform. Indicate the amplitude, offset, and period of the waveform.

141

4. Calculate the magnitude of the input, Vg, at 120Hz, 360Hz, and 600Hz, using the Fourier series for the square wave. Record in the table below.

Frequency	Calculated Mag. Vg Volts-peak		Measured Mag. Vo Volts-peak	Measured Angle Vo Degrees
120				
360				
600				

5. Set the function generator to produce a sine wave at each frequency of the same magnitude as calculated in step 4 above. Measure and record the peak magnitude and phase angle of the output, Vo, at each frequency.

Analysis

1. Calculate the transfer function of the filter as a phasor for DC, 120Hz, 360Hz, and 600Hz.

2. Calculate the output of the filter at each frequency using phasor analysis.

3. Compare the magnitude of the measured output, Vo, with the calculated output at each frequency. Calculate the percent difference between the measured and calculated results.

4. Compare the measured phase angle of the output, Vo, with the calculated phase angle of the output at each frequency. Calculate the percent difference between the measured and calculated results.

5. Simulate the circuit. Use the FFT feature to display the magnitude of the first four terms of the output of the filter and compare the simulated results to the calculated results.

    Note: Set the time domain display to about 8 cycles. Change the frequency axis of the FFT to display 0 to 800Hz. Simulated output magnitudes in LTspice are in volts-rms.

LTspice Simulation: FFT

1. Draw the circuit using your part values. Set the voltage source to "PULSE" and the analysis to transient as shown in the schematic on the right. The example shown here is for a 100Hz, 2V p-p square wave. Set up the voltage source using the frequency and magnitude used in the experiment.

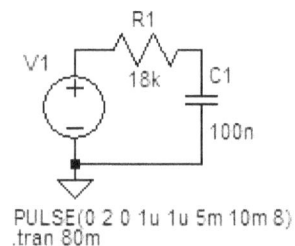

PULSE(0 2 0 1u 1u 5m 10m 8)
.tran 80m

2. Select "Run". Click on plot window that opens. Click on "View" in main menu and select FFT.

3. Click on the vertical axis of the plot to change the scale to linear. The plot will show the rms voltage values of the harmonics.

4. Click on the horizontal axis of the plot to change the frequency range to 1KHz.

The simulation results below show the first five components of the spectrum of the source, V(n001), and the output of the filter, V(n002). Note that the DC component of V(n001) is not shown.

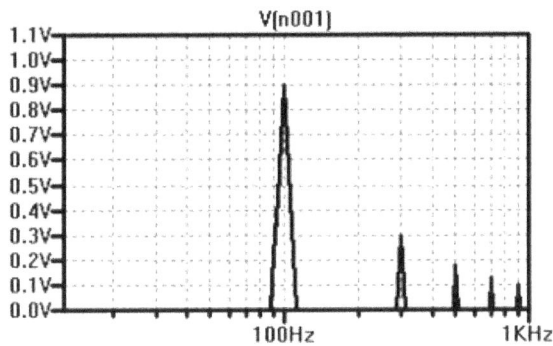

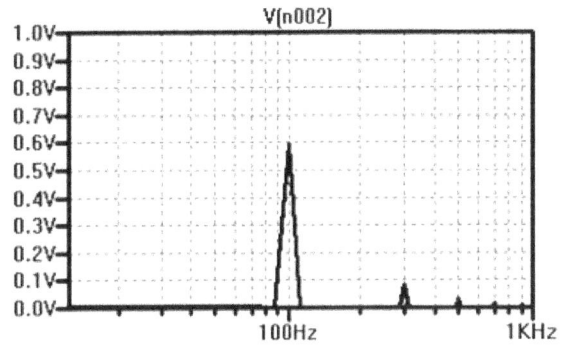

History Note:

Jean Baptiste Joseph Fourier (21 March 1768 – 16 May 1830) A French mathematician and physicist known for initiating the investigation of Fourier series and their applications to problems of heat transfer and vibrations. The Fourier transform and Fourier's Law are also named in his honor. Fourier is also generally credited with the discovery of the greenhouse effect.

From Wikipedia, the free encyclopedia

Experiment 44: Band Pass Filter / FFT / Square Wave

Introduction

According to Fourier analysis, a square wave may be considered to be a superposition of an infinite number of odd harmonic frequencies whose amplitudes decrease inversely with frequency. The fundamental is the lowest frequency of the square wave. The Fourier series of a square wave voltage, whose peak-to-peak amplitude is V, and average value is V_{AVE}, may be expressed as:

$$v(t) = V_{AVE} + \frac{2V}{\pi} \sum_{n}^{\infty} \frac{1}{n} \sin(2\pi n f_0 t), \text{ where n is an odd}$$

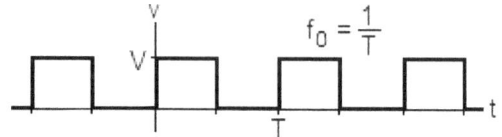

integer and the fundamental frequency is f_0.

If a square wave is applied to a band-pass filter, the wave's frequency components will be affected by the filter's frequency response. The filter's output frequency spectrum will be different than the input frequency spectrum.

In Fourier analysis, the concepts of "time domain" and "frequency domain" are important to understand. The square wave input to the filter and output from the filter are considered to be expressed in the time domain when its amplitude is expressed as a function of time. An oscilloscope normally plots waveforms in the time domain. When the amplitude of the waveform is expressed as a function of frequency, it is expressed in the frequency domain.

Using Fourier analysis, it is possible to convert time domain expressions into the frequency domain, and frequency domain expressions into the time domain. This may be done by a computer algorithm called the "Fast Fourier Transform", or "FFT", which is built into many analysis software packages, including *PSpice* and *LTspice*. The objective of this experiment is to investigate the relationship between time domain analysis and frequency domain analysis. Acquired time domain data will be compared to simulation data using *OrCAD PSpice*.

Although the Fourier series of any time domain waveform consists of an infinite number of harmonic frequencies, an adequate approximation may be obtained using only the first several terms of the series. This exercise will involve the average value of the series plus the first 3 harmonics. These are:

$$v(t) = V_{AVE} + \frac{2V}{\pi} \sin(2\pi f_0 t) + \frac{2V}{3\pi} \sin(6\pi f_0 t) + \frac{2V}{5\pi} \sin(10\pi f_0 t)$$

Procedure

Equipment and Parts
Function Generator, Digital Oscilloscope with FFT, and Breadboard Resistors: 3.3K, 18K, ¼ watt, 5% Capacitors: 1µF (non-polarized), 100nF, 5%.

1. Connect the circuit on the right. Connect the oscilloscope channel 1 to node N1 and Channel 2 to node N2.

2. Vi is the function generator set to produce a 20Hz, 2V peak-to-peak, square wave with a 1 volt offset (goes between 0 and 2V).

3. Display 10 cycles of the input and output waveforms as shown below.

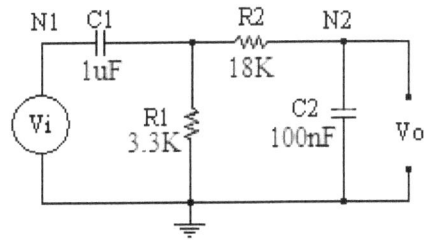

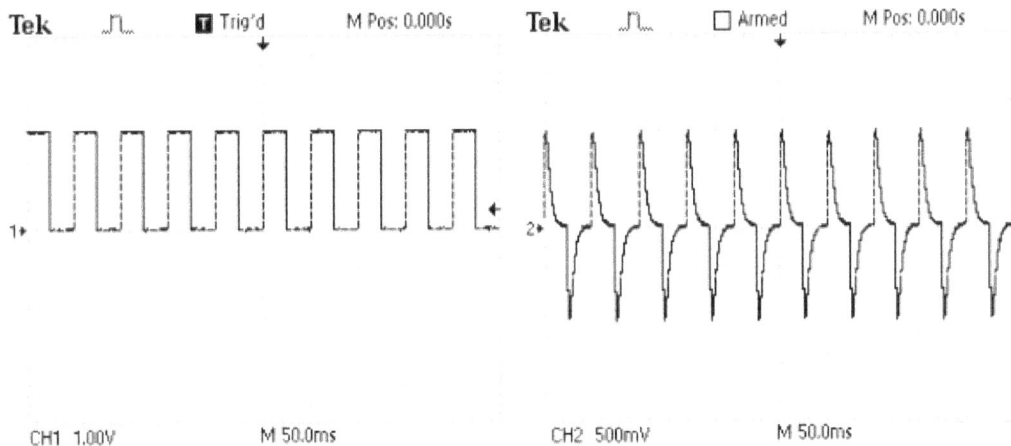

It may be helpful to read the section on displaying the Fourier spectrum of your instrument's user manual.

The TDS1002 uses a "Math FFT" algorithm to generate the Fourier spectrum from the acquired time domain data. The time domain waveform needs to be set up carefully to get an accurate display of the spectrum. Please note the following:

a. The FFT spectrum is calculated from the center 2048 points of the time domain waveform. Since there are 2500 points in the 10 divisions of the entire display, the FFT is calculated from about the center 8 divisions.

b. There is a tradeoff between frequency resolution and the bandwidth of the displayed spectrum (due to aliasing).

c. The highest frequency (and bandwidth) that can be measured accurately by a digitizing oscilloscope is one half the sample rate (the Nyquist frequency).

d. The TDS1002 transforms 2048 time domain points to 1024 frequency domain points resulting in a spectrum whose bandwidth is equal to the Nyquist frequency.

e. The amplitude is displayed in dB where 0 dB is equal to 1 V rms.

$$dB_{VRMS} = 20\log\frac{V_{RMS}}{1Vrms} \ .$$

f. Use the cursors to determine the amplitudes of the harmonics.

g. Use data capture such as the TekXL toolbar for greater accuracy

145

4. To obtain the FFT spectrum of channel 1: Push the "MATH MENU" button, select "FFT", set the MATH FFT source to channel 1.

   Note the resulting display on the right. The frequency scale is 250Hz/Div. The amplitude scale is in dB, where 0dB = 1V rms.

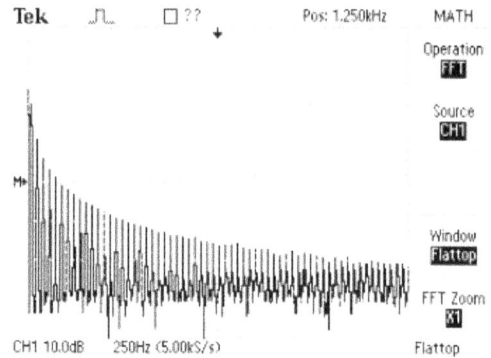

5. The frequency scale needs to be expanded by a factor of 10 in order to accurately measure the frequency components. This could be done with the time base (sec/div) control. However, this would reduce the bandwidth from 2500Hz to 250Hz. The TDS1002 has an FFT zoom control that provides a zoom up to 10. A zoom factor of 10 was used to obtain the input (CH1) and output (CH2) frequency spectrums shown below.

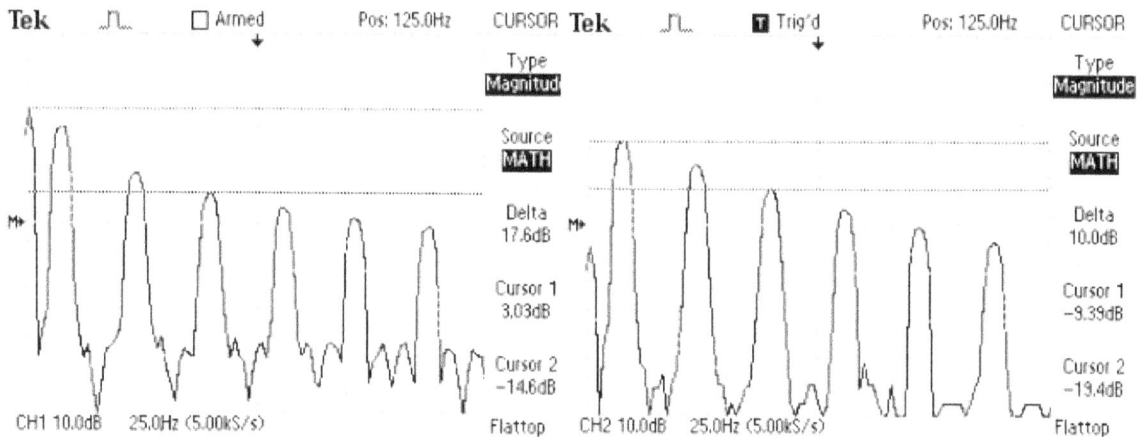

Cursors should be used to measure the amplitudes. The input frequency spectrum (channel 1) above shows an rms amplitude value of 3dB at 0Hz. This corresponds to 1.414V rms, which is equal to a 1V average value (DC offset). The fifth harmonic at 100Hz shows an amplitude of −14.6dB, corresponding to a peak voltage of 0.254V. The output frequency spectrum shows the fifth harmonic amplitude to be −19.4dB.

6. Display the first six harmonics of the input and output spectra of your band pass filter. Record the amplitude of the DC component and the first three harmonics of the input and output spectra (DC, 20Hz, 60Hz, 100Hz).

   Vi_{DC}: _____dBV Vi_{20}: _____dBV Vi_{60}: _____dBV Vi_{100}: _____dBV

   Vo_{DC}: _____dBV Vo_{20}: _____dBV Vo_{60}: _____dBV Vo_{100}: _____dBV

Analysis

Note: Use a spreadsheet as applicable to do the analysis below.

1. The Fourier series of the input waveform is that of a 2V peak-to-peak amplitude square wave with a 1.0V DC component (average value). Determine the theoretical magnitude of the waveform's DC component, fundamental, third harmonic and fifth harmonic. Convert your calculated results to dBV using the following equation:

$$dBV = 20 \log \frac{Vrms}{1Vrms}.$$

Compare your measured values to the theoretical values by calculating the percent difference between them.

2. Simulate the filter circuit and use the FFT feature to obtain the frequency spectrum of the output. Compare the simulated results to your calculated results from step 1 above.

3. Simulate the circuit and use the frequency sweep analysis to obtain a Bode Plot of the filter's response (similar to plot shown below). Compare the results to analysis steps 1 and 2 above.

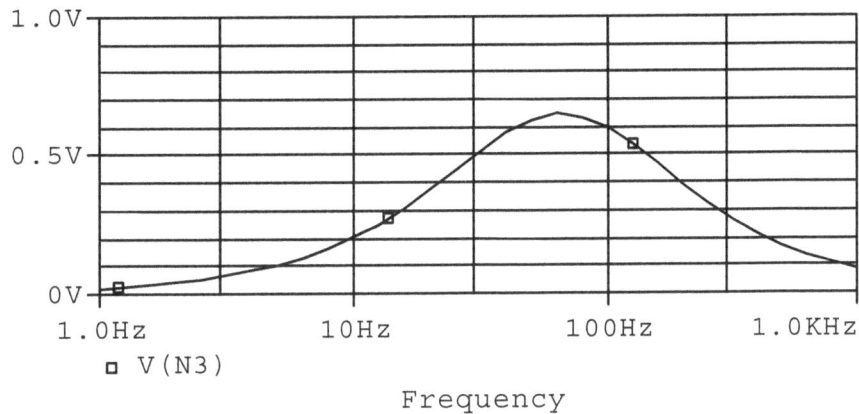

Experiment 45: Band Pass Filter / FFT / Triangle Wave

Introduction

According to Fourier analysis, a triangle wave may be considered to be a superposition of an infinite number of odd harmonic frequencies whose amplitudes decrease inversely with frequency. The fundamental frequency is the lowest frequency of the triangle wave. The Fourier series of a triangle wave voltage, whose peak-to-peak amplitude is V, and average value is V_{AVE}, may be expressed as:

$$v(t) = V_{AVE} - \frac{4\,V}{\pi^2} \sum_{n}^{\infty} \frac{1}{n^2} \cos(2\pi n\, f_0 t),$$

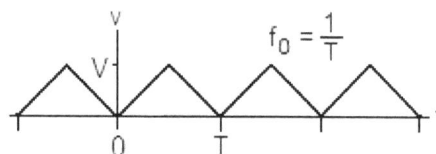

where n = odd, and the fundamental frequency is f_0.

If a triangle wave is applied to a band-pass filter, the wave's frequency components will be affected by the filter's frequency response. The filter's output frequency spectrum will be different than the input spectrum.

Compared to the square wave, the amplitude of the harmonics of a triangle wave decrease much more rapidly with frequency. The Fourier series for the average plus the first 3 terms is:

$$v(t) = V_{AVE} + \frac{4\,V}{\pi^2} \cos(2\pi f_0 t) + \frac{4V}{9\pi^2} \cos(6\pi f_0 t) + \frac{4V}{25\pi^2} \cos(10\pi f_0 t).$$

Procedure

Equipment and Parts

Function Generator, Digital Oscilloscope with FFT, and Breadboard
Resistors: 3.3KΩ, 18KΩ, ¼ watt, 5%
Capacitors: 1μF (non-polarized), 100nF, 5%.

---

1. Connect the circuit on the right. Connect the oscilloscope channel 1 to node N1 and channel 2 to node N2.

2. Vi is the function generator. Set it to produce a 20Hz, 2V peak-to-peak, triangle wave with a 1 volt offset (goes between 0 and 2V).

3. Set the oscilloscope to display 10 cycles of the input and output waveforms.

4. Set the oscilloscope to FFT (refer to the settings in last experiment if you are using a *Tek TDS* oscilloscope). Display the first six harmonics of the input and output spectra of your band pass filter.

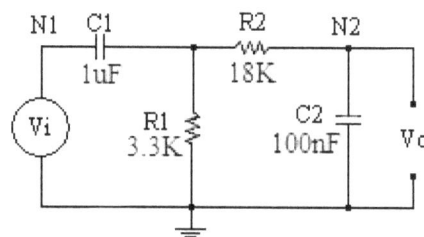

Record the amplitude of the DC component and the first three harmonics of the input and output spectra (DC, 20Hz, 60Hz, 100Hz).

Vi_{DC}: _____dB Vi_{20}: _____dB Vi_{60}: _____dB Vi_{100}: _____dB

Vo_{DC}: _____dB Vo_{20}: _____dB Vo_{60}: _____dB Vo_{100}: _____dB

Analysis

Note: Use a spreadsheet as applicable to do the analysis below.

1. The Fourier series of the input waveform is that of a 2V peak-to-peak amplitude triangle wave with a 1.0V DC component (average value). Determine the theoretical magnitude of the waveform's DC component, fundamental, third harmonic and fifth harmonic. Convert your calculated results to dBV using the following equation:

$$dBV = 20 \log \frac{Vrms}{1Vrms}.$$

Compare your measured values to the theoretical values by calculating the percent difference between them.

2. Simulate the filter circuit and use the FFT feature to obtain the frequency spectrum of the output (refer to LTspice example in experiment 43). Compare the simulated results to your calculated results from step 1 above.

Appendix 1: Electric Circuits Lab Report Information

Introduction

The purpose of the laboratory is three fold. First is to learn how to use typical laboratory instruments to make a variety of measurements. Second is to reinforce the lecture material by demonstrating practical applications of electric circuit theory. Third is to learn how to analyze data and write a lab report.

Each laboratory experiment will involve several, and possibly all, of the following skills. You should be able to use a word processor and a spreadsheet such as Microsoft Word and Excel. You will learn all the other skills listed below.

1. How to connect electrical circuits using a circuit diagram.
2. How to measure voltage, current, and resistance.
3. How to collect and record data.
4. How to present data in lab a report.
5. How to simulate electric circuits using circuit simulation software.

The Lab Station

A typical electronic lab station consists of the following instruments: power supply, digital multi-meter, function generator, frequency counter, oscilloscope, computer, and computer data acquisition equipment and software.

A solder-less breadboard, or proto typing board, is used to connect the circuits. The instruments are connected to the breadboard as needed. **As a general rule, it is best to use the smallest number of wires, and the smallest number of connections, when doing an experiment. This reduces the number of things that can go wrong, and it is easier to see how everything is connected.**

Many people must use the laboratory and its equipment, so it is important that everyone using the lab cooperates in keeping it in good condition. At the end of each lab period you are responsible for straightening up your lab station.

Make sure that all the equipment is turned off, unless told otherwise. Remove all loose papers and disposable items (loose wires, paper clips, etc.). Test equipment leads should be restored to their proper positions. Report any equipment malfunctions or other problems to your instructor or the lab assistant.

Grading

Specific grading information will be provided by your instructor. The typical requirements of an electric circuits lab course includes a lab notebook, and at least one formal lab report. A lab exam may also be given. Typical formats for lab notebooks and a formal lab reports are presented on the following pages.

Lab Notebook Procedures

Laboratory experiments should be recorded in a notebook which is square-ruled with approximately five lines per inch. The notebooks should be bound (not loose leaf or spiral). They should be orderly and complete.

Data from the experiment should be recorded directly into the notebook. It is a good idea to prepare the data tables before coming to the lab. It may be acceptable to use pencil; however, many organizations require that a pen be used for official notebooks. If a computer printout or other addendum is specifically required to be attached to your notebook, it should be thoroughly taped on all edges or glued securely in the book. Do not staple in your book.

The laboratory notebook is a record or log of what happened in the laboratory. The significance of any measurement, calculation, or observation made during the course of the experiment may not be apparent until the experiment is completed. Therefore, the lab notebook should contain a record of all the measurements and observations.

Each experiment should include the following:

1. The experiment number and title.
2. Statement of purpose.
3. Schematics and/or drawings.
4. Data tables for recording experimental results.
5. Graphs, curves or additional tables that may be needed.
6. Analysis and conclusions.

You will be using computer spreadsheets to analyze and organize your data and results. Do not record your measurements directly into the spreadsheet program. Copy your measurements from your lab notebook into the spreadsheet. Keep your notebook as neat and well organized as possible.

In some cases, you may need specific information about the test instruments. For example, you may need to know the internal resistance of a measuring instrument, or the specific settings of the instrument. Be sure to record all of the information that you may need.

Specific lab notebook procedures will be provided by the lab instructor.

Formal Lab Report

At least one written formal report may be required. The instructor will specify which experiment or experiments may require a formal report. The general format for the formal report is given below. Each report should organized into clearly defined and labeled sections which should include the following:

1. Abstract
2. Introduction or problem statement.
3. Experimental setup and procedure.
4. Results and discussion.
5. Conclusions.

The exact organization depends on the experiment. Some experiments may have several parts, and each part may require a separate introduction, procedure, and results section. In some cases, an appendix may be needed. Additional information and suggestions on the content of each section is presented below.

1. *Abstract:* A short summary of the purpose, methods, and results of the experiment. Often just one paragraph is needed. It is usually written last (after you have a good idea of what the experiment was about and what actually happened).

2. *Introduction or problem statement:* A clearly stated explanation of the experiment. State the purpose of the experiment.

3. *Experimental setup and procedure:* Circuit diagrams, equipment used, measurement methods, drawings, etc., as needed.

4. *Results and discussion:* A tabulation of the measurements made with tables and graphs, as needed. A comparison to theoretical expectation which may include calculations and computer simulations. A statistical analysis, percent deviation of the results from the theoretical, which may include graphs and a spreadsheet analysis. A discussion of the results, including possible causes of errors, etc.

5. Conclusions: A short summary of the results, significance of the results, comments on the results. It usually takes experience to write a good conclusion.

You may want to include an appendix containing references cited in the report, such as the manufacturers data sheet on parts used, or information on the instruments used in the experiment. A typical formal lab report may be between 5 and 10 pages long and include diagrams, graphs, and tables. It must be done using a word processor.

Diagrams, tables and graphs should be imported directly into the word processor, using for example, the "clipboard" in Microsoft Windows. Being able to use software, such as word processors, spreadsheets, and engineering design and simulation tools is becoming more and more important.

Appendix 2: 3-Phase Voltage Sources

The 3-phase experiments in this lab book are intended to aide a student's understanding of 3-phase circuits. Any low voltage 3-phase source may be used that can supply pure 60Hz sine waves with an amplitude of 12 volts peak-to-peak per phase and current of 25mA rms per phase.

A simple phase tripler circuit is described here that a student may build on a breadboard. It uses power op-amps and phase shift networks to convert a 60Hz sine wave input to 3-phase sine wave output. It also requires a ±12VDC power supply (about 100mA).

This circuit is also available in a breadboard plug-in version and a printed circuit board version. Both can be obtained in kit form or in assembled and calibrated form..

Description

This phase tripler circuit converts single-phase 60Hz ac to three-phase 60Hz AC using just two simple phase shift networks. The single-phase source may be a function generator or the ac voltage from a step down transformer.

The maximum input amplitude is 18V peak-to-peak. The output amplitude is the same as the input. The diagram on the left below shows three-phase ac produced by three sources connected in a 4-wire wye configuration. A block diagram of the phase tripler circuit is shown below on the right.

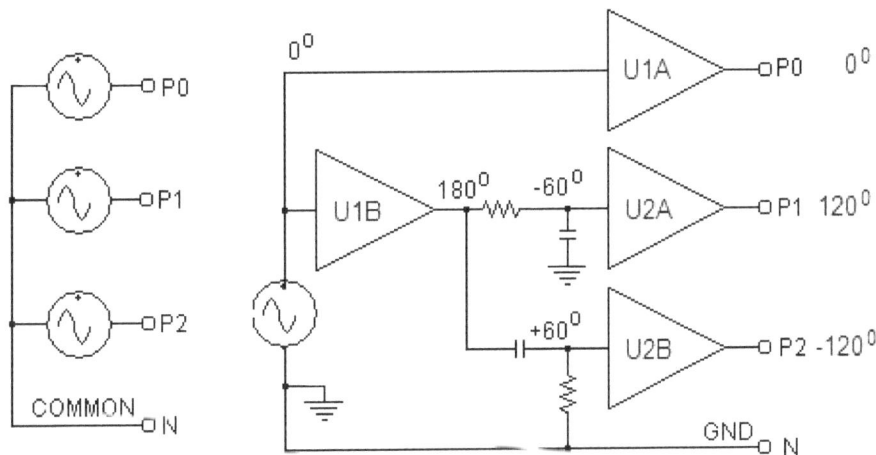

The zero degree reference source is buffered by the op-amp, U1A, and output at P0. An op-amp phase inverter, U1B, shifts the phase of the input by 180 degrees. This phase shifted voltage is applied to a 60 degree lag network to obtain a net phase shift of 120 degrees, and to a 60 degree lead network to obtain a net phase shift of 240 (-120) degrees.

The magnitude of the voltage output by these networks is one half of the input, so op-amps, U2A and U2B, are used to amplify the voltage by a factor of two. The 120 degree and 240 degree voltages are output at P1 and P2 with the same amplitude as the voltage at output P0. Trimmer potentiometers are used to adjust the phase and amplitudes of the outputs.

This is a wye-connected source with a common connection that is also the circuit ground for the power supply and op-amps. Instrument grounds cannot be connected to the output of an op-amp because that would cause a short circuit.

Measurements should be made with respect to the common ground, unless the measuring instrument is known to be isolated from ground.

The recommended maximum load current for each output is 25mA. At 18V peak-to-peak, this corresponds to 160mW per output.

Construction

This circuit may be built on a breadboard, or a printed circuit board can be obtained from www.zapstudio.com.

The *Phase Tripler II*, the breadboard plug-in version, is shown on the right. Its Pin connections are given below.

Pin 1: Generator   Pin 2: Ground

Pin 4: +12VDC   Pin 6: -12VDC

Pin 9: P0   Pin 10: P1   Pin 11: P2

Pin 12: Neutral (ground).

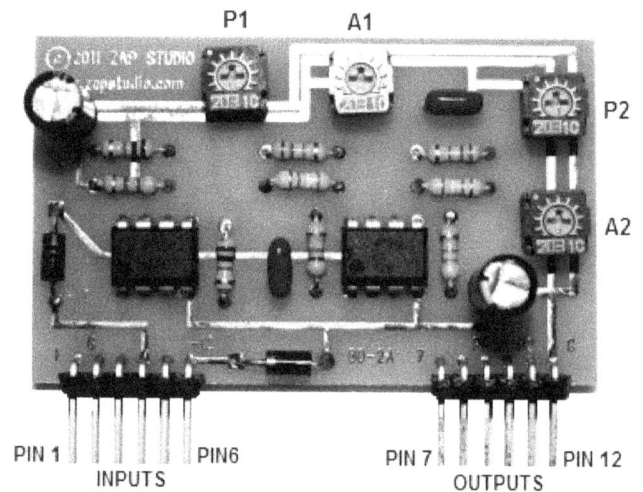

The circuit board layout for the *Phase Tripler*, the pcb version, is shown on the right.

Both circuit boards use the same circuit and components. The calibration instructions on the next page apply equally to the *Phase Tripler*, *Phase Tripler II*, and the breadboard connected version.

P0 = P000   P1 = P120   P2 = P240

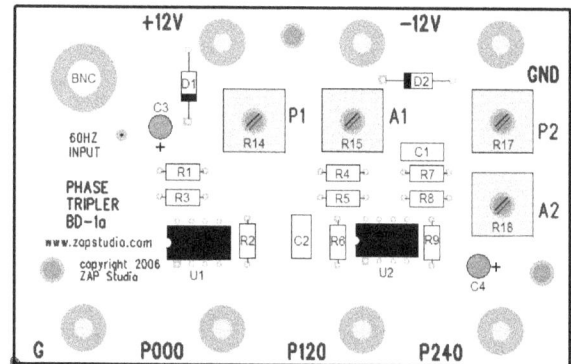

The circuit is designed to operate at 60Hz only. Other frequencies will not produce the correct phase angles. 60Hz was chosen to correspond to the power distribution frequency in the USA.

It is also possible to use a 117VAC to 6.3VAC step-down transformer as a signal source instead of a function generator. But the transformer waveform may be somewhat distorted and noisy. A 60Hz band-pass filter could be used to improve the wave form.

Calibration Procedure

1. Connect a function generator to the input and set it to produce a 12V peak-to-peak, 60Hz, sine wave.

2. Connect the oscilloscope channel 1 to output P0. The output at P0 should be exactly 12V peak-to-peak at 60Hz. Set the trigger to channel 1. Connect channel 2 of the oscilloscope to output P1.

3. Set the oscilloscope time base to 2ms per division. Center both traces. Adjust pot P1 so that the positive slope P1 zero crossing is exactly 11.11ms after the positive slope P0 zero crossing.

   Adjust pot A1 so that the amplitude of P1 output is exactly 12V peak-to-peak.

4. Connect the oscilloscope channel 2 to output P2. Set the oscilloscope time base to 2ms per division. Center both traces. Adjust pot P2 so that the positive slope P2 zero crossing is exactly 5.56ms after the positive slope P0 zero crossing.

   Adjust pot A2 so that the amplitude of P2 output is exactly 12V peak-to-peak.

The graphs on the right show a rotating phasor diagram and the resulting sinusoids.

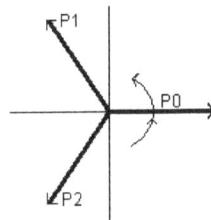

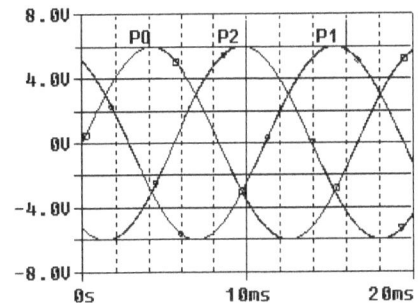

Circuit Schematic Diagram

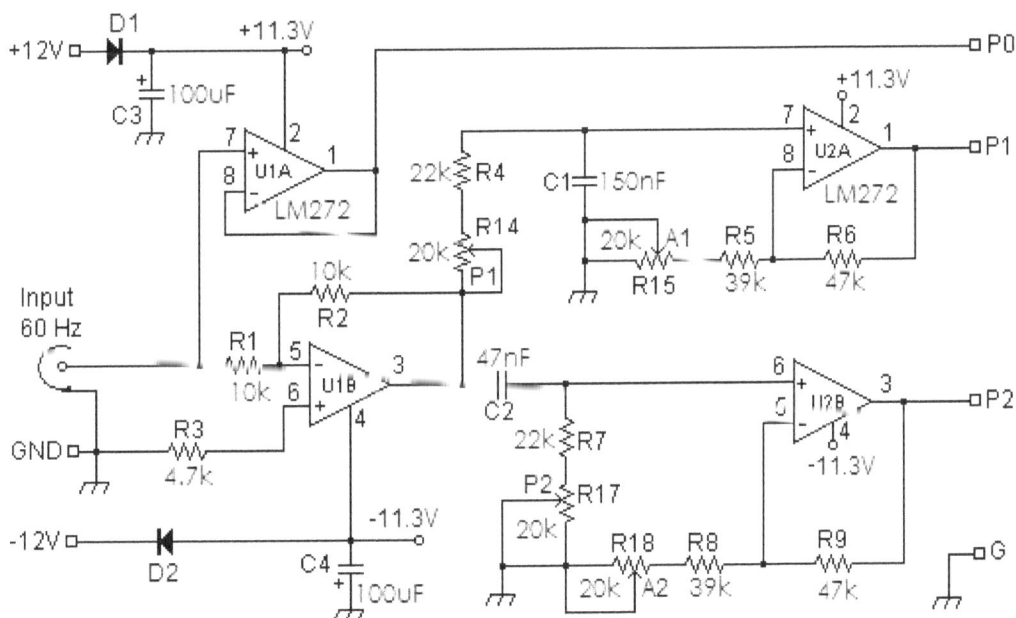

Parts List

U1, U2: L272M (Mouser 512-L272M) power op-amps.

R14, R15, R17, R18: 20K trim-pots.

R1, R2: 10K, ¼, 5%.   R3: 4.7K, ¼ watt, 5%.   R4, R7: 22K, ¼ watt, 5%.

R5, R8: 39K, ¼ watt, 5%.   R6, R9: 47K, ¼ watt, 5%.

C1: 150nF, 5%.   C2: 47nF, 5%   C3, C4: 100µF.

D1, D2: 1N4001 diodes.

Input and output connectors depend on which pc board is used. The Phase Tripler is supplied with a BNC connector and banana binding posts. The Phase Tripler II is supplied with two six pin headers. If the circuit is assembled on a breadboard, input and output connectors are not required.

The *Phase Tripler* and *Phase Tripler II* are available in kit form and in assembled and calibrated form from ZAP Studio LLC.   www.zapstudio.com

Appendix 3: DC Motor–Generator Information

Description

Two small DC motors are used in this motor-generator set that are connected together by tygon tubing. A gear wheel and a photo-interrupter switch are used to make a tachometer. The physical layout of the system is shown below.

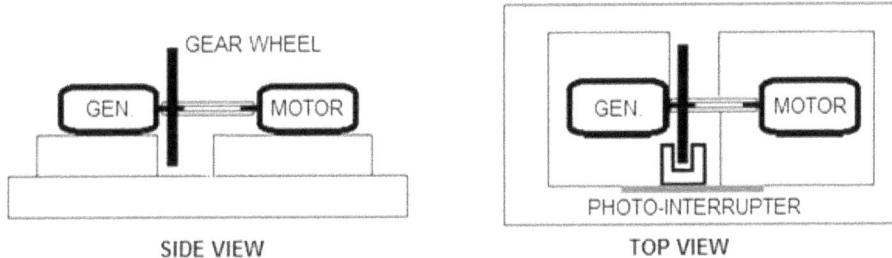

SIDE VIEW   TOP VIEW

The motors may be mounted on pine boards, although a hardwood like oak may be better. They can be clamped to the boards with plastic electrical wire clamps.

The connecting tygon tubing fits tightly over the motor shafts. It has a 1/16th inch inside diameter and is cut to a 1.5 inch length (if the motors are too close together frictional effects may reduce the motor's efficiency). It is also important that the motor shafts are able to move in and out a little, about one or two millimeters, to reduce frictional effects.

The gear wheel's center hole should be drilled to a diameter which provides a tight fit over the tygon tubing and motor shaft, as shown in the diagram above. The photo-interrupter is mounted on a board which is attached to the motor mounting boards as shown in the diagram above on the right.

A Motor-generator and photo-interrupter wiring diagram is shown on the right. The photo-interrupter circuit's supply voltage, Vs, may range from 6V to 12V.

A 1Ω resistor in series with the motor is used to measure the motor current. The magnitude of the voltage, Vi, in volts is equal to the magnitude of the motor current in amps.

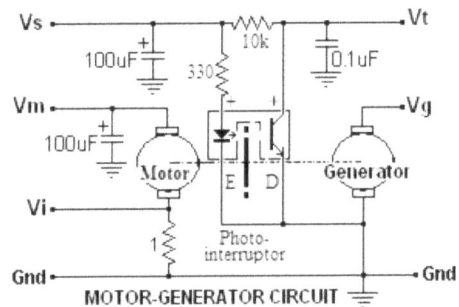

MOTOR-GENERATOR CIRCUIT

Parts:
Photo-interrupter: OPB 815L, Newark Electronics.
DC Motor: 3-9V,6800RPM,0.25A, 231781, Jameco.
Gear: 26 pitch, 40 teeth, 3 holes, 162093, Jameco.

The picture on the right shows an assembled motor-generator set with a small breadboard and additional parts. These boards are used at Portland Community College for electric circuits and feedback control system experiments.

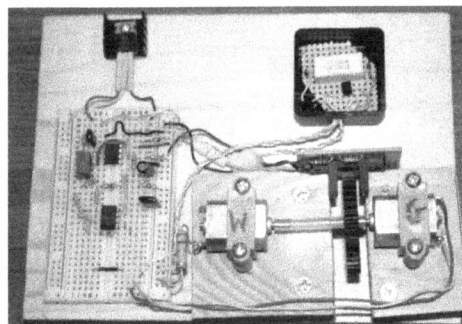

Appendix 4: Flashlight Circuit Information

For convenience, the flashlight circuit can be assembled on a small board and the circuit nodes can be labeled as shown below.

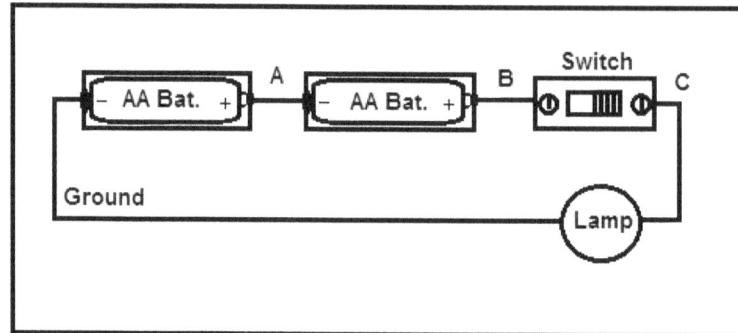

Appendix 5: LTspice Information

LTspice IV is free to download from Linear Technology:
http://www.linear.com/designtools/software/

For very *useful and quick* reference guide to the most common LTspice tasks, check this PDF out:
http://pages.suddenlink.net/wa5bdu/ltguide.pdf

For more detailed assistance, you can download the Getting Started Guide:
http://ltspice.linear.com/software/LTspiceGettingStartedGuide.pdf